METHODS IN MOLECULAR BIOLOGY

Series Editor
John M. Walker
School of Life and Medical Sciences
University of Hertfordshire
Hatfield, Hertfordshire, AL10 9AB, UK

For further volumes:
http://www.springer.com/series/7651

The Golgi Complex

Methods and Protocols

Edited by

William J. Brown

Department of Molecular Biology and Genetics, Cornell University, Ithaca, NY, USA

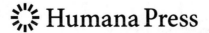

Editor
William J. Brown
Department of Molecular Biology and Genetics
Cornell University
Ithaca, NY, USA

Videos can also be accessed at http://link.springer.com/book/10.1007/978-1-4939-6463-5_1.

ISSN 1064-3745   ISSN 1940-6029 (electronic)
Methods in Molecular Biology
ISBN 978-1-4939-8208-0   ISBN 978-1-4939-6463-5 (eBook)
DOI 10.1007/978-1-4939-6463-5

Preface

The Golgi complex has fascinated cell biologists for decades because of its unique architecture and role as a "Grand Central Station" for trafficking soluble and membranous cargo through the secretory and endocytic pathways. Indeed, the Golgi's unusual architecture itself has thwarted efforts to fully understand how secretory cargo is imported, passed through, and exported from the Golgi complex. Nevertheless, tremendous progress has been made to reveal the molecular mechanism underlying the functional organization of the Golgi complex. Accordingly, new methods and techniques have been developed to address long-standing questions about the Golgi complex, which are the focus of this book. The chapters herein encompass a diverse set of methods for studying the Golgi complex, ranging from live and fixed cell imaging techniques to in vitro biochemical reconstitution systems. The book is targeted to both recent, i.e., new students, and established investigators in the field to provide strong practical instructions that can be directly applied to their research programs. Each chapter provides a detailed set of specific instructions, which should enable anyone to successfully complete the assays. Moreover, this methods series is particularly helpful by encouraging contributors to include a series of "Notes," which often contain subtle nuances or unappreciated facts that often get omitted from straight methods protocols.

Ithaca, NY, USA   *William J. Brown*

Contents

Contributors

INMACULADA AYALA • *Institute of Protein Biochemistry, National Research Council of Italy, Naples, Italy*

COLLIN BACHERT • *Department of Biological Sciences, Carnegie Mellon University, Pittsburgh, PA, USA*

MARIE E. BECHLER • *MRC Centre for Regenerative Medicine, The University of Edinburgh, Edinburgh, UK*

MARIANO BISBAL • *Laboratorio Neurobiología, Instituto Investigación Médica Mercedes y Martín Ferreyra (INIMEC-CONICET), Córdoba, Argentina; Universidad Nacional de Córdoba, Córdoba, Argentina; Instituto Universitario Ciencias Biomédicas Córdoba, Córdoba, Argentina*

JESSICA BAILEY BLACKBURN • *Department of Physiology, University of Arkansas for Medical Sciences, Little Rock, AR, USA*

JUAN S. BONIFACINO • *Cell Biology and Neurobiology Branch, Eunice Kennedy Shriver National Institute of Child Health and Human Development, National Institutes of Health, Bethesda, MD, USA*

DYLAN J. BRITT • *Cell Biology and Neurobiology Branch, Eunice Kennedy Shriver National Institute of Child Health and Human Development, National Institutes of Health, Bethesda, MD, USA*

WILLIAM J. BROWN • *Department of Molecular Biology and Genetics, Cornell University, Ithaca, NY, USA*

ALFREDO CACERES • *Laboratorio Neurobiología, Instituto Investigación Médica Mercedes y Martín Ferreyra (INIMEC-CONICET), Córdoba, Argentina; Universidad Nacional de Córdoba, Córdoba, Argentina; Instituto Universitario Ciencias Biomédicas Córdoba, Córdoba, Argentina*

EDWARD B. CLUETT • *Department of Biology, Ithaca College, Ithaca, NY, USA*

ANTONINO COLANZI • *Institute of Protein Biochemistry, National Research Council of Italy, Naples, Italy*

KASEY J. DAY • *Department of Molecular Genetics and Cell Biology, University of Chicago, Chicago, IL, USA*

BERIT EBERT • *ARC Centre of Excellence in Plant Cell Walls, School of BioSciences, The University of Melbourne, Parkville, VIC, Australia*

HOLGER ERFLE • *BioQuant, University of Heidelberg, Heidelberg, Germany*

GINNY G. FARÍAS • *Cell Biology and Neurobiology Branch, Eunice Kennedy Shriver National Institute of Child Health and Human Development, National Institutes of Health, Bethesda, MD, USA*

PAUL DE FIGUEIREDO • *Department of Microbial Pathogenesis and Immunology, Texas A&M Health Science Center, Bryan, TX, USA; Department of Veterinary Pathobiology, Texas A&M University, TX, USA; Norman Borlaug Center, Texas A&M University, TX, USA*

J. CHRISTOPHER FROMME • *Department of Molecular Biology and Genetics, Weill Institute for Cell and Molecular Biology, Cornell University, Ithaca, NY, USA*

BENJAMIN S. GLICK • *Department of Molecular Genetics and Cell Biology, University of Chicago, Chicago, IL, USA*

MANUEL GUNKEL • *BioQuant, University of Heidelberg, Heidelberg, Germany*

SARA FASMER HANSEN • *Department of Plant and Environmental Sciences, Faculty of Science, University of Copenhagen, Frederiksberg, Denmark; Joint BioEnergy Institute and Physical Biosciences Division, Lawrence Berkeley National Laboratory, Berkeley, CA, USA*

JOSHUA L. HEAZLEWOOD • *Joint BioEnergy Institute and Physical Biosciences Division, Lawrence Berkeley National Laboratory, Berkeley, CA, USA; ARC Centre of Excellence in Plant Cell Walls, School of BioSciences, The University of Melbourne, Parkville, VIC, Australia*

VICTOR W. HSU • *Division of Rheumatology, Immunology, and Allergy, Brigham and Women's Hospital, Boston, MA, USA; Department of Medicine, Harvard Medical School, Boston, MA, USA; Brigham and Women's Hospital, Boston, MA, USA*

BYUNG-HO KANG • *School of Life Sciences, Centre for Cell and Developmental Biology and State Key Laboratory of Agrobiotechnology, Chinese University of Hong Kong, Hong Kong, China*

ADAM D. LINSTEDT • *Department of Biological Sciences, Carnegie Mellon University, Pittsburgh, PA, USA*

ALBERTO LUINI • *Institute of Protein Biochemistry, National Research Council, Naples, Italy; Instituto di Ricovero e Cura a Carattere Scientifico SDN, Naples, Italy*

VLADIMIR V. LUPASHIN • *Department of Physiology, University of Arkansas for Medical Sciences, Little Rock, AR, USA*

DOROTA MASZCZAK-SENECZKO • *Laboratory of Biochemistry, Faculty of Biotechnology, University of Wroclaw, Wrocław, Poland*

TERESA OLCZAK • *Laboratory of Biochemistry, Faculty of Biotechnology, University of Wroclaw, Wrocław, Poland*

MARIUSZ OLCZAK • *Laboratory of Biochemistry, Faculty of Biotechnology, University of Wroclaw, Wrocław, Poland*

JON E. PACZKOWSKI • *Department of Molecular Biology and Genetics, Weill Institute for Cell and Molecular Biology, Cornell University, Ithaca, NY, USA; Department of Molecular Biology and Howard Hughes Medical Institute, Princeton University, Princeton, NJ, USA*

EFFROSYNI PAPANIKOU • *Department of Molecular Genetics and Cell Biology, University of Chicago, Chicago, IL, USA*

SEUNG-YEOL PARK • *Division of Rheumatology, Immunology, and Allergy, Brigham and Women's Hospital, Boston, MA, USA; Department of Medicine, Harvard Medical School, Boston, MA, USA*

GONZALO QUASSOLLO • *Laboratorio Neurobiología, Instituto Investigación Médica Mercedes y Martín Ferreyra (INIMEC-CONICET), Córdoba, Argentina; Universidad Nacional de Córdoba, Córdoba, Argentina; Instituto Universitario Ciencias Biomédicas Córdoba, Córdoba, Argentina*

CARSTEN RAUTENGARTEN • *ARC Centre of Excellence in Plant Cell Walls, School of BioSciences, The University of Melbourne, Parkville, VIC, Australia*

RICCARDO RIZZO • *Institute of Protein Biochemistry, National Research Council, Naples, Italy*

JOHN A. SCHMIDT • *Department of Biology, Villanova University, Villanova, PA, USA*

LINA SONG • *Department of Biological Sciences, Carnegie Mellon University, Pittsburgh, PA, USA*

PAULINA SOSICKA • *Laboratory of Biochemistry, Faculty of Biotechnology, University of Wroclaw, Wroclaw, Poland*

VYTAUTE STARKUVIENE • *BioQuant, University of Heidelberg, Heidelberg, Germany; Department of Biochemistry and Molecular Biology, Faculty of Natural Sciences, Joint Life Sciences Center, University of Vilnius, Vilnius, Lithuania*

VINCENT S. TAGLIABRACCI • *Department of Molecular Biology, University of Texas Southwestern Medical Center, Dallas, TX, USA*

KEVIN D. THORSEN • *Department of Molecular Biology and Genetics, Cornell University, Ithaca, NY, USA*

JIANZHONG WEN • *Discovery Bioanalytics, Merck and Co, Rahway, NJ, USA*

JUNYU XIAO • *State Key Laboratory of Protein and Plant Gene Research, School of Life Sciences, Peking University, Beijing, China; Peking-Tsinghua Center for Life Sciences, Peking University, Beijing, China*

JIA-SHU YANG • *Division of Rheumatology, Immunology, and Allergy, Brigham and Women's Hospital, Boston, MA, USA; Department of Medicine, Harvard Medical School, Boston, MA, USA*

Chapter 1

4D Confocal Imaging of Yeast Organelles

Kasey J. Day, Effrosyni Papanikou, and Benjamin S. Glick

Abstract

Yeast cells are well suited to visualizing organelles by 4D confocal microscopy. Typically, one or more cellular compartments are labeled with a fluorescent protein or dye, and a stack of confocal sections spanning the entire cell volume is captured every few seconds. Under appropriate conditions, organelle dynamics can be observed for many minutes with only limited photobleaching. Images are captured at a relatively low signal-to-noise ratio and are subsequently processed to generate movies that can be analyzed and quantified. Here, we describe methods for acquiring and processing 4D data using conventional scanning confocal microscopy.

Key words Yeast, Confocal, 4D microscopy, Photobleaching, Deconvolution, ImageJ

1 Introduction

Live-cell fluorescence microscopy provides crucial information about organelle dynamics. Ideally, an entire 3D cell volume (Z-stack) is captured at each time point, yielding a 4D data set that allows intracellular structures to be tracked for many minutes. In the case of yeast cells, 4D imaging is facilitated by the small size of the cells and the relatively low copy numbers of some compartments. A variety of methods have been described for yeast 4D imaging, including widefield microscopy with a specialized deconvolution algorithm [1] and spinning-disk confocal microscopy with a custom high-sensitivity system [2]. Here, we describe our preferred method using conventional laser-scanning confocal microscopy. This approach uses readily available instruments and software.

The major challenge in 4D imaging is to minimize photobleaching and the accompanying photodamage to the cells. Yeast organelles are frequently tagged with fluorescent proteins such as

Electronic supplementary material: The online version of this chapter (doi:10.1007/978-1-4939-6463-5_1) contains supplementary material, which is available to authorized users. Videos can also be accessed at http://link.springer.com/chapter/10.1007/978-1-4939-6463-5_1.

William J. Brown (ed.), *The Golgi Complex: Methods and Protocols*, Methods in Molecular Biology, vol. 1496,
DOI 10.1007/978-1-4939-6463-5_1, © Springer Science+Business Media New York 2016

GFP and mCherry. These fluorophores, as well as endocytic tracers such as the dye FM 4-64 [3], are prone to bleaching by confocal lasers. A single Z-stack of a yeast cell typically comprises at least 20 optical slices, and a 4D movie typically involves capturing a Z-stack every few seconds for 5–15 min, so the total acquisition can run to thousands of individual confocal sections. Preservation of fluorescence signals under these conditions requires careful attention to multiple parameters.

To minimize photobleaching, the laser power should be as low as possible. We find that photobleaching rates are nonlinear with respect to laser intensity. During multicolor imaging, a laser used to excite one fluorophore can also bleach a second fluorophore—e.g., a 488-nm laser used to excite GFP can also bleach mCherry—so reducing the laser power can have dramatic benefits. In addition, the dwell time of the laser should be low. To meet these criteria, we use the fastest available scan rates and then perform line averaging to obtain usable signals.

Pixel size as determined by the Zoom setting is a crucial parameter. To image a fluorescent structure in a manner that avoids information loss, the sampling interval should be no higher than the Nyquist limit, which is ~2.3 times smaller than the resolution of the optical system [4]. Yeast imaging generally employs a 1.4-NA oil immersion objective, giving a resolution on the order of 200 nm, so the pixel size should be <90 nm. In practice, oversampling down to a pixel size about half of the Nyquist limit tends to improve image quality, but this luxury is usually unavailable for 4D imaging because reducing the pixel size results in slower scans and more photobleaching. As a compromise, we use a pixel size of 65–70 nm, or ~85 nm for weak fluorescence signals.

Similarly, the Z-step interval between confocal sections must be small enough to avoid information loss, but large enough to keep photobleaching low and to enable rapid acquisition of a Z-stack. We use a Z-step interval of ~0.25 μm, or ~0.35 μm for weak fluorescence signals. Enough confocal sections should be used to ensure that fluorescent structures throughout the cell volume will be fully captured, even if the sample shows drift along the Z-axis during movie acquisition. We usually collect 20–30 optical slices per Z-stack.

The signal-to-noise ratio for fluorescent organelles can be enhanced in several ways. A simple tactic is to overexpress the tagged protein of interest. This approach is sometimes unavoidable, in which case control experiments can be performed to confirm that overexpression has not substantially changed the biological system [5]. Whenever possible, a protein should be expressed at endogenous levels by adding a tag through gene replacement at the chromosomal locus. For both overexpression and endogenous gene tagging, we routinely use cassettes encoding 3× or 6× tandem copies of a fluorescent protein [6]. The fluorescence signal scales with the copy number of the fluorescent protein [7]. In addition, we minimize the background noise from the cell by using completely nonfluorescent minimal media devoid of riboflavin and folic acid [8].

Even when the fluorescence signal is strong and clean as viewed by widefield microscopy, the confocal sections captured under conditions suitable for 4D imaging can look alarmingly noisy. The number of photons per pixel is usually very low, so labeled structures may not look smooth and may be embedded in a background that includes shot noise. It is important to accept relatively noisy images at the collection stage with the expectation that the noise can be reduced through image processing. A simple way to reduce shot noise and smooth the true signal is with a 3×3 hybrid median filter, which can be applied once or iteratively [9]. A more sophisticated approach is deconvolution with commercial software. Standard deconvolution methods are not ideal for noisy fluorescence images and may erase weak signals [1]. Until better deconvolution methods become available, we recommend that if the signal-to-noise ratio is low, the data should be preprocessed with a hybrid median filter before deconvolution. In general, empirical testing is needed to devise an image processing routine that adequately preserves the desired signal while removing most of the noise.

After filtering and/or deconvolution, it may be useful to correct for photobleaching. For most fluorophores, the bleaching rate approximates an exponential decay, and a 4D data set can be corrected for exponential photobleaching using an ImageJ plugin. This correction makes it easier to view and quantify the final movies. However, the information content of the Z-stacks diminishes with photobleaching, so if the photobleaching is severe at later time points, little or no signal will be recovered. Reducing the laser power may be beneficial because even though the early images will be of somewhat lower quality, the usable signal will persist for significantly longer.

A 4D movie can be viewed using a volume renderer such as the commercial Imaris or Volocity software, but a simpler and often sufficient alternative is to project each Z-stack to create a 2D image. The projections can then be assembled into a movie. Although maximum intensity projections are often used for convenience, they exaggerate weakly labeled structures, so average projections are preferable for quantitation [9]. A disadvantage of projections is that structures located at different depths in the cell volume may appear to be close together or merged. To address this issue, a labeled structure of interest can be tracked in the original 4D data set, and an edited movie can be generated by manually erasing the other fluorescence signals and projecting only the signal from the labeled structure [5, 10]. Such edited movies are ideal for quantifying the dynamics of individual labeled organelles.

As an example, Fig. 1 shows the first frames from Videos 1 and 2, which are 4D confocal movies of a yeast strain expressing a

Fig. 1 Illustration of the effect of image processing on 4D movies. Shown are the first frames from Videos 1 and 2, which are 4D movies from the same data set before and after deconvolution, respectively. Yeast cells expressing GFP-Vrg4 as an early Golgi marker and Sec7-DsRed as a late Golgi marker were imaged by confocal microscopy, with cell images in the *blue* channel. Z-stacks of 24 optical sections each were collected every 2 s. Where indicated, the *red* and *green* channels were deconvolved. The movie frames are average projections of the Z-stacks. In the deconvolved movie, maturation events can be observed when *green* Golgi cisternae turn *red* [5]

GFP-tagged early Golgi protein and a DsRed-tagged late Golgi protein [5]. These markers gave relatively strong signals. In the left panel of Fig. 1 and in Video 1, the raw data were average projected to illustrate the noise level in the images. In the right panel of Fig. 1 and in Video 2, the deconvolved data were average projected to illustrate how noisy images can be processed to obtain usable movies.

2   Materials

2.1   Instrumentation and Software

1. *Microscope*: Leica SP5 or comparable laser scanning confocal microscope with inverted optics, a fast scanner, and high sensitivity detectors.

2. *Objective lens*: Plan-Apo 100× or 63× lens with an NA of 1.4 or greater. The magnification is less important than the NA and lens quality.

3. *Lasers*: Multiple laser lines, minimally including a 488-nm laser for exciting green fluorophores and a 561-nm laser for exciting red fluorophores.

4. *Stage adaptor*: Standard adaptor for holding dishes with cover glass bottoms.

5. *Z-step motor*: Piezoelectric stepper motor for rapid and accurate collection of Z-stacks.

6. *Image manipulation and analysis software*: ImageJ freeware obtained from http://imagej.nih.gov/ij/. Custom ImageJ plugins for 4D data analysis can be found in the supplemental

material for a recent paper [10], and an ImageJ plugin that corrects for photobleaching can be obtained from http://cmci.embl.de/downloads/bleach_corrector.

7. *Deconvolution software*: Huygens Essential (Scientific Volume Imaging).

8. *Movie processing software*: QuickTime Player 7 (Apple).

2.2 Materials

1. Confocal imaging is performed with yeast strains that have been either engineered to express one or more fluorescently tagged marker proteins, or incubated with a fluorescent tracer dye. For expressing tagged proteins, we typically modify the endogenous chromosomal copy of a gene by clean gene replacement using the pop-in/pop-out method [11, 12]. If desired, a tagged gene can be overexpressed by integrating a vector containing a strong promoter [5]. These approaches have the advantage that the cells do not carry free plasmids, so the expression level within the culture is uniform and the strain is stable even when grown in nonselective medium. However, in some cases it is useful to express a tagged gene on a centromeric plasmid, because cell-to-cell variability in expression enables the choice of cells with an appropriate level of fluorescence.

2. Yeast cells are grown with shaking in baffled flasks either in SD dropout medium consisting of 0.67% yeast nitrogen base with ammonium sulfate, 2% glucose, and CSM complete supplement mixture or dropout mixture (Sunrise Science Products), or in a nonfluorescent SD medium (NSD) in which the yeast nitrogen base is replaced with a mixture of salts and vitamins lacking riboflavin and folic acid (*see* **Note 1**). For a strain that does not require selection to maintain a plasmid, a preculture is grown in rich YPD medium (1% yeast extract, 2% peptone, 2% glucose) and stored at 4 °C, and this preculture is used to inoculate a culture in SD or NSD medium at a dilution of 1:1000 to 1:5000 for overnight growth to logarithmic phase (optical density at 600 nm of about 0.4–0.8).

3. Cover glass-bottom 35-mm dishes with 14-mm microwells and No. 1.5 thickness are obtained from Bioptechs or MatTek, optimally with high tolerance glass of 170 ± 5 μm.

4. Concanavalin A is dissolved in water to 2 mg/mL, and stored at 4 °C for up to a week.

3 Methods

3.1 Preparing the Cells

1. Grow one or more yeast strains overnight to logarithmic phase in 5 mL NSD medium in a 50-mL baffled flask.

2. Prepare a cover glass bottom dish: pipet 250 μL concanavalin A into the dish to submerge the cover glass, wait 15 min, wash thoroughly with ddH$_2$O, and let dry.

3. Just before imaging, adhere cells to the cover glass: pipet 250 μL from the culture into the dish to submerge the cover glass, wait 10 min, and rinse gently several times with NSD by pipetting. Finally, add 2–3 mL of fresh NSD to the dish.

4. If treatment with a drug or dye is required, replace the NSD with fresh medium containing the drug or dye. If appropriate, remove the medium and replace it with fresh medium before imaging.

3.2 Confocal Imaging

1. Configure the microscope to use a 100× or 63× high-NA oil immersion lens. Ensure that the DIC prism is not in the light path, or the fluorescence images will be sheared and distorted. In the case of a 100× objective with a Leica STED system, the quarter wave plate should be in place or the red and green signals will be offset along the Z-axis.

2. Set the frame size to approximately 256×128 (*see* **Note 2**).

3. Set the Zoom to a level that will give a pixel size of 65–70 nm, or up to ~85 nm if the signal is weak.

4. Set the scanner to operate at maximum speed. Use bidirectional mode, as long as the brightfield image confirms that the images from the two directions can be accurately aligned.

5. Set the pinhole to 1.2 Airy units (*see* **Note 3**).

6. Set the line averaging at 4–8 (*see* **Note 4**).

7. Choose a Z-step size of ~0.25 μm, or up to ~0.35 μm if the signal is weak. Note the directionality of Z-stack acquisition (toward or away from the coverslip). To avoid confusion, be consistent with this setting between movies.

8. Set an appropriate time interval between Z-stacks. An interval of 2 s is typically suitable, but shorter intervals may be needed to track very dynamic compartments, and somewhat longer intervals may be preferable if the signal is weak.

9. Set the lasers to the lowest levels that will generate workable signals (3–10% on the SP5) (*see* **Note 5**). For fluorescence data, use the highest sensitivity detectors (HyD on the SP5). For brightfield data, a less sensitive detector is adequate (PMT on the SP5). Set the detector gains to appropriate levels (400–500 for the HyD and 300–350 for the PMT on the SP5).

10. Choose emission windows that will maximize the information collected while minimizing bleedthrough between channels. For visualization of GFP that is excited with a 488-nm laser, the emission window is 495–550 nm, and for visualization of mCherry that is excited with a 561-nm laser, the emission window is 575–750 nm (*see* **Note 6**).

11. Store each confocal section as an 8-bit RGB image using the microscope vendor's standard file format, with the brightfield view of the cells stored in the blue channel.

3.3 Filtering

For weak signals, deconvolution often erases the structures of interest, leading to flickering movies that are impossible to analyze. In this case, it may help considerably to preprocess the data with a hybrid median filter followed by a Gaussian blur.

1. In ImageJ, open the 4D data file as a TIFF hyperstack. Then use Image > Color > Split Channels to separate the channels. With the custom plugin called "Filter Hybrid Median", filter the individual optical slices in the fluorescence channels using a single iteration of the 3D hybrid median filter. If this filter removes too much information, try a single iteration of the standard 2D hybrid median filter (*see* **Note 7**).

2. After hybrid median filtering, use Process > Filters > Gaussian Blur to do a 2D Gaussian blur with a 1-pixel radius for the optical slices in the fluorescence channels. Then use Image > Color > Merge Channels to merge the channels once again into a hyperstack.

3.4 Deconvolution and Bleach Correction

Although the relatively large pixels and Z-steps used for 4D imaging are not optimal for deconvolution, this processing method is useful for cleaning up noisy images and generating smooth objects that are suitable for downstream analysis. We use the Huygens algorithm, but other deconvolution algorithms should give similar results. As described above, when the signal is weak, the Huygens software may erase a structure unless the data are preprocessed by filtering. This effect can reportedly be attributed to the mathematical form of the deconvolution algorithm, which is not ideal for noisy fluorescence data [1].

1. If the 4D data file was produced directly by the confocal microscope software, the Huygens interface should provide an option for reading the imaging parameters directly. Alternatively, if the 4D data set was preprocessed in ImageJ to generate a TIFF file, these parameters will need to be entered manually. Ensure that the deconvolution software has the correct setting for the directionality of Z-stack acquisition. The refractive index of the embedding medium, which is a yeast cell, can be estimated as 1.35–1.40.

2. Deconvolve the fluorescence channels but not the brightfield channel. A signal-to-noise ratio of 10 is usually appropriate. Other parameters such as background subtraction can be varied in empirical tests until the output appears to be an accurate rendering of the original fluorescence signals.

3. Save the output from the deconvolution as an 8-bit TIFF image sequence.

4. In ImageJ, choose File > Import > Image Sequence to convert the image sequence to a stack. Then choose Image > Hyperstacks > Stack to Hyperstack and input the

appropriate order for the confocal sections (e.g., "xyzct"), number of channels, slices per stack, and frames.

5. To correct for photobleaching (*see* **Note 8**), first use Image > Color > Split Channels to separate the channels. For each of the fluorescence channels, use Plugins > EMBLtools > Bleach Correction, and choose "Exponential Fit (Frame-wise)". Make sure the plot shows a good curve fit, then close the plot window and log window and original channel window, leaving the new corrected channel window. Finally, use Image > Color > Merge channels to regenerate a hyperstack with the bleach-corrected fluorescence data.

3.5 Editing 4D Data Sets with Custom ImageJ Plugins

These manipulations with custom ImageJ plugins [10] are used when the goal is to create an edited 4D data set that includes only one or a few structures. The edited data set can be analyzed to quantify the time course of the fluorescence signals.

1. Examine a projected movie that was generated as described in **Section 3.6** from a non-edited 4D data set. Identify candidate structures that can potentially be tracked for an extended period without interference from other nearby structures, keeping in mind that the 4D data set may permit resolution of structures that occasionally overlap in a projection.

2. Identify the 8-bit TIFF hyperstack that was used to generate the non-edited projected movie. With the custom plugin called "Make Montage Series", open this hyperstack, scale the images, and create the montage series.

3. With the custom plugin called "Edit Montage Series", choose a single structure that will be tracked through part or all of the movie. Delete the fluorescence signals for all of the other structures at each time point.

4. With the custom plugin called "Montage Series to Hyperstack", generate a hyperstack for the edited 4D data set. The images in this hyperstack will normally be magnified due to scaling that occurred during creation of the montage series.

5. With the custom plugin called "Analyze Edited Movie", open the hyperstack for the edited 4D data set and specify the time interval between Z-stacks. The "Red" and "Green" columns in the output file represent the total fluorescence signals from the structure of interest at each time point. Save this file, and open it in Microsoft Excel or another data analysis program to plot the fluorescence signals as a function of time.

3.6 Processing 4D Data Sets for Presentation

After performing any or all of the filtering, deconvolution, bleach correction, and editing steps, a 4D data set needs to be converted from a hyperstack to a form that is suitable for presentation and publication.

1. The first step is to convert the data to 16-bit, to prevent information loss during the subsequent average projection. In ImageJ, choose Image > Color > Split Channels. For each of the resulting three windows, choose Image > Type > 16-bit. Then choose Image > Color > Merge Channels to restore the hyperstack. Finally, choose Process > Math > Multiply, and use a multiplication factor of 256 (*see* **Note 9**).

2. If desired, merge the fluorescence from two 4D movies using the custom ImageJ plugin called "Merge Two Hyperstacks". This plugin allows two edited 4D movies to be merged, or alternatively allows one 4D movie to be placed above the other—e.g., an original movie above the corresponding edited movie.

3. The next step is to create an average projection. In ImageJ, choose Image > Stacks > Z Project. If desired, select a subset of each Z-stack by choosing the Start and Stop slices. Choose the "Average Intensity" option.

4. Adjust the brightness and contrast of the individual channels as follows. Choose Image > Adjust > Brightness/Contrast. For each channel, press the "Auto" button. These settings can be fine tuned using the sliders or the "Set" button.

5. To add a time stamp, use Image > Stacks > Label. Use the format 00:00, with the appropriate time interval in seconds. To place the label in the lower left corner, set the Y location to be 1 less than the Y value of the lowest row of pixels in the image.

6. The final step is to make a movie. In ImageJ, choose File > Save As > AVI. Choose PNG compression. A frame rate of 10 fps is generally suitable. If desired, open the resulting AVI file with QuickTime Player 7 and then export a MOV file, which will often have a much smaller file size with no loss of image quality.

4   Notes

1. Nonfluorescent glucose medium (NSD) can be made by assembling the components of normal SD medium except for riboflavin and folic acid. To make 100 mL of a 500× vitamin stock solution, add 100 mg calcium pantothenate, 500 mg myo-inositol, 20 mg niacin, 10 mg *p*-aminobenzoic acid, 20 mg pyridoxine hydrochloride, 20 mg thiamine hydrochloride, and 20 mg biotin, and store at 4 °C. To make a 500× stock solution of cobalt chloride, add 0.1 g of cobalt chloride hexahydrate per liter, and store at 4 °C. To make 1 L of NSD, add 20 g glucose, 5 g ammonium sulfate, 5 g potassium phosphate monobasic, 1 g magnesium sulfate heptahydrate, 0.5 g sodium chloride, and 0.1 g calcium chloride dihydrate. Add

2 mL of the vitamin stock solution and 2 mL of the cobalt chloride stock solution. Add YNB trace elements (ANACHEM) from a concentrated stock solution, which is stored at 4 °C. If desired, add a CSM complete supplement mixture or drop-out mixture. Adjust the pH to 5.5 with NaOH.

2. Frame height is the main determinant of the total scan time. If desired, a wider frame can be used to capture more data.

3. The typical recommendation is to use 1.0 Airy unit. However, that setting assumes optimal conditions with no refractive index mismatch between the sample and the oil/glass, whereas the actual refractive index of the cell is lower than that of the oil/glass. Empirically, we find that increasing the pinhole to 1.2 Airy units results in the capture of significantly more light with a negligible decrease in resolution.

4. Fast scans with line averaging tend to cause less photobleaching than slower scans without line averaging. Image quality is dramatically improved by line averaging, but at the expense of increasing the acquisition time. A line averaging setting of 4–8 is typically a good compromise.

5. Each laser should be used at a power that yields a decent signal while keeping the photobleaching rate acceptably low. During collection, the signal often is barely visible and is accompanied by significant noise, but such data sets can be processed to obtain usable movies. Set the laser power to 5% or even lower if possible. Photobleaching increases nonlinearly with laser intensity, so a reduction in laser power may yield a signal that is weaker at first but persists for much longer.

6. Ensure that the confocal microscope is configured with notch filters to suppress background signals from the excitation lasers.

7. Hybrid median filters work optimally when the image collection settings are near the Nyquist limit. For imaging with a high-NA objective, the pixel size should be in the range of 60–90 nm and the Z-stack interval should be in the range of 0.2–0.4 μm. Filtering before deconvolution is not generally encouraged by makers of deconvolution software because it alters the characteristics of the data, but empirically, this approach sometimes improves the movies by preserving biologically meaningful signals.

8. Unless the signal is quite strong, photobleaching is usually significant. The correction procedure involves fitting the signal intensity decay to an exponential curve and then multiplying each Z-stack by an appropriate factor to compensate. This procedure is effective up to a point, but as the original signal becomes fainter, the weaker signals in the data set will be progressively lost and will not be recovered after the correction for photobleaching.

9. After conversion to 16-bit and multiplication by 256, the image display may not be scaled appropriately, but the data are still suitable for further processing.

References

1. Arigovindan M, Fung JC, Elnatan D, Menella V, Chan YH, Pollard M, Branlund E, Sedat JW, Agard DA (2013) High-resolution restoration of 3D structures from widefield images with extreme low signal-to-noise-ratio. Proc Natl Acad Sci U S A 110:17344–17349

2. Kurokawa K, Ishii M, Suda Y, Ichihara A, Nakano A (2013) Live cell visualization of Golgi membrane dynamics by super-resolution confocal live imaging microscopy. Methods Cell Biol 118:235–242

3. Vida TA, Emr SD (1995) A new vital stain for visualizing vacuolar membrane dynamics and endocytosis in yeast. J Cell Biol 128:779–792

4. Pawley JB (2006) Handbook of biological confocal microscopy, 3rd edn. Springer, New York

5. Losev E, Reinke CA, Jellen J, Strongin DE, Bevis BJ, Glick BS (2006) Golgi maturation visualized in living yeast. Nature 22: 1002–1006

6. Genové G, Glick BS, Barth AL (2005) Brighter reporter genes from multimerized fluorescent proteins. Biotechniques 39:814–822

7. Connerly PL, Esaki M, Montegna EA, Strongin DE, Levi S, Soderholm J, Glick BS (2005) Sec16 is a determinant of transitional ER organization. Curr Biol 15:1439–1447

8. Bevis BJ, Glick BS (2002) Rapidly maturing variants of the *Discosoma* red fluorescent protein (DsRed). Nat Biotechnol 20:83–87

9. Hammond AT, Glick BS (2000) Raising the speed limits for 4D fluorescence microscopy. Traffic 1:935–940

10. Papanikou E, Day KJ, Austin JI, Glick BS (2015) COPI selectively drives maturation of the early Golgi. Elife 4:13232. doi:10.7554/eLife

11. Rossanese OW, Soderholm J, Bevis BJ, Sears IB, O'Connor J, Williamson EK, Glick BS (1999) Golgi structure correlates with transitional endoplasmic reticulum organization in *Pichia pastoris* and *Saccharomyces cerevisiae*. J Cell Biol 145:69–81

12. Rothstein R (1991) Targeting, disruption, replacement, and allele rescue: integrative DNA transformation in yeast. Methods Enzymol 194:281–301

Chapter 2

Imaging the Polarized Sorting of Proteins from the Golgi Complex in Live Neurons

Ginny G. Farías, Dylan J. Britt, and Juan S. Bonifacino

Abstract

The study of polarized protein trafficking in live neurons is critical for understanding neuronal structure and function. Given the complex anatomy of neurons and the numerous trafficking pathways that are active in them, however, visualization of specific vesicle populations leaving the Golgi complex presents unique challenges. Indeed, several approaches used in non-polarized cells, and even in polarized epithelial cells, have been less successful in neurons. Here, we describe an adaptation of the recently developed Retention Using Selective Hooks (RUSH) system (Boncompain et al., Nat Methods 9:493–498, 2012), previously used in non-polarized cells, to analyze the polarized sorting of proteins from the Golgi complex to dendrites and axons in live neurons. The RUSH system involves the retention of a fluorescently tagged cargo protein fused to the streptavidin-binding peptide (SBP) in the endoplasmic reticulum (ER) through the expression of an ER-hook protein fused to streptavidin. Upon D-biotin addition, the cargo protein is released and its traffic to dendrites and axons can be analyzed in live neurons.

Key words Hippocampal neurons, Neuron dissection, Neuronal culture, Neuron transfection, Live-cell imaging, Polarized trafficking, Axon initial segment, RUSH system, Trafficking from the Golgi complex, Kymograph analysis

1 Introduction

Neurons are polarized cells with a highly asymmetric structure. They comprise axonal and somatodendritic domains with different compositions of proteins in both the cytoplasm and plasma membrane. Both domains also have distinct subdomains, including presynaptic terminals at axon tips and postsynaptic terminals in dendrites [1–3]. The asymmetry of protein localization across these domains reflects their specialized functions and allows the vectorial transmission of information from the presynaptic terminals of one neuron to the postsynaptic terminals of another [1–3]. Maintenance of polarity and dynamic response to developmental and environmental inputs are important to proper neuronal function [1–3]. Thus, the means by which polarized sorting occurs, as

William J. Brown (ed.), *The Golgi Complex: Methods and Protocols*, Methods in Molecular Biology, vol. 1496, DOI 10.1007/978-1-4939-6463-5_2, © Springer Science+Business Media New York 2016

well as the native role of each protein in its intended domain, are of mechanistic interest both in vivo and in vitro.

A common method for studying neuronal structure and function is the primary culture of hippocampal neurons from embryonic rats or mice. When maintained in culture, neurons progress through a characteristic sequence of developmental stages [4]. Within 2 days of plating, cells begin to establish polarity, and the one neurite destined to become the axon grows to several times the length of the others. The first synapses begin to form after 3–4 days in culture [4]. Around day in vitro (DIV) 5, a key event in neuronal polarization occurs with the accumulation of proteins such as ankyrin G (AnkG) at the axon initial segment (AIS) [3]. Among other roles, the AIS proteins play a major part in neuronal polarity, serving as a barrier to plasma membrane diffusion between the axonal and somatodendritic domains [3]. Axonal growth continues beyond this stage, producing dense networks of polarized, interconnected cells as additional synapses form over the following days [4].

The ability to transport proteins to their correct intracellular locations is critical for achieving neuronal polarity, and cells employ a variety of mechanisms to maintain this composition [2, 5, 6]. As in non-neuronal cells, newly synthesized proteins transferred from the endoplasmic reticulum (ER) to the Golgi complex undergo biosynthetic sorting [1]. The Golgi in neurons takes on a unique structure consisting of a somatic Golgi stack, which is preferentially oriented toward the longest dendrite in cultured neurons, as well as Golgi outposts present in dendrites but excluded from the axon (Fig. 1) [7]. Upon arrival at the *trans*-most compartment of the Golgi complex, the *trans*-Golgi network (TGN), proteins undergo sorting into specific transport carriers depending on their intended destinations, after which polarized transport mechanisms move these carriers into either axonal or somatodendritic domains [1]. Various machineries responsible for this sorting have been characterized, chief among them the adaptor protein (AP) complexes and the coat protein clathrin [1, 6]. Following budding from the TGN, proteins and their transport carriers may be subject to additional sorting and fusion events. For example, proteins may undergo active endocytosis in one domain but retention in another [5]. Certain proteins reach their destinations via transcytosis and thus traffic to one domain before being endocytosed, moving once again through the soma, and entering the opposite domain [5]. Proteins may also be targeted to lysosomes for degradation in a domain-specific manner [8].

In studying the initial sorting and ultimate fate of transmembrane proteins, direct visualization of trafficking from the Golgi complex is a useful approach. It was in neurons that Camillo Golgi first observed and described the apparatus that bears his name [9], and the work of recent decades has established trafficking from the

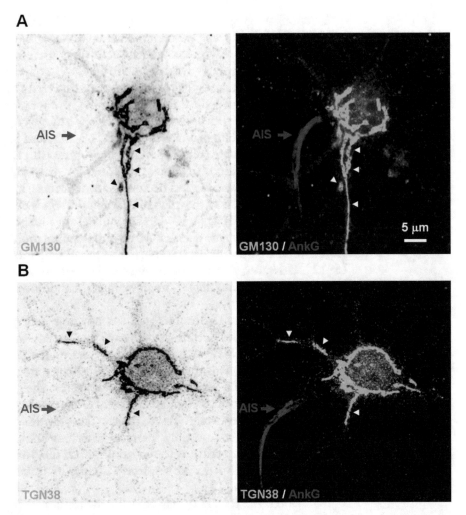

Fig. 1 Golgi complex in hippocampal neurons. (**a, b**) Immunofluorescence of neurons on DIV 10 stained for the Golgi marker GM130 (**a**) or the *trans*-Golgi network (TGN) marker TGN38 (**b**), together with the AIS marker AnkG. GM130 and TGN38 staining is shown in negative grayscale and *green* in the merged images, and AnkG is shown in *red* in the merged images. *Red arrows* point to the AIS. Scale bar, 5 μm. Notice the localization of the Golgi complex to the soma, and dendrites as indicated by *arrowheads*

Golgi complex as a fundamental aspect of neuronal development and function [1, 6]. Knowledge of the mechanisms and dynamics of protein sorting is important in understanding normal cellular processes and neuropathological disorders [2]; however, the study of sorting in live neurons has been challenging because proteins may be simultaneously present in the biosynthetic, endosomal, and lysosomal pathways at steady state [5]. Photobleaching and photoconvertible reporter expression have been widely used to study plasma membrane insertion of transmembrane proteins in neurons [10–12], but these methods are less useful when imaging the sorting of proteins from the Golgi complex.

In contrast to neurons, in non-polarized cells and even in polarized epithelial cells, which contain apical and basolateral domains, several approaches have been successfully used to dissect and image the sorting of proteins from the Golgi complex. For example, previous studies have used temperature reduction to 19 °C to retain proteins in the TGN [13, 14], the temperature-sensitive mutant of the Vesicular Stomatitis Virus glycoprotein (VSVG tsO45) to retain proteins in the ER [15–17], and pharmacologically controlled aggregation of proteins in the ER [18]. The most powerful and extensively used procedure for analyzing traffic in the biosynthetic pathway has been cargo synchronization using the VSVG transport assay, in which the mutant protein is misfolded and retained in the ER at 40 °C. Upon temperature shift to 32 °C, the protein folds and moves as a synchronous population to the Golgi complex before being transported to the plasma membrane [15–17]. However, this system has been of limited use in neurons because of their susceptibility to unphysiological conditions such as changes in temperature.

Recently, an alternative method for imaging synchronous transport through the secretory pathway, the Retention Using Selective Hooks (RUSH) system, was developed by Franck Perez and colleagues [19]. This system consists of the expression of two fusion proteins from a bicistronic expression vector: the "hook," stably expressed in a donor compartment and fused to a core streptavidin, and the "reporter," fused to the Streptavidin-Binding Peptide (SBP), which reversibly interacts with the hook. Upon reversion of the interaction through addition of the vitamin D-biotin to the cell, the reporter can freely resume its traffic in the biosynthetic pathway [19].

We adapted the RUSH system to study the trafficking of transmembrane proteins from the Golgi complex to dendrites and axon in live neurons. Here we describe a detailed protocol for the preparation of hippocampal neurons, transfection of cDNAs including those of the RUSH system, staining of endogenous proteins to identify the somatodendritic and axonal domains, live-cell imaging using the RUSH system, and kymograph analysis for particle transport in dendrites and axons.

2 Materials

2.1 Neuron Preparation and Transfection

1. Ultrapure water, cell culture grade.
2. High-precision glass coverslips, 18-mm diameter, No. 1.5H (Marienfeld).
3. 70% nitric acid.
4. Borate buffer (0.1 M, pH 8.5, 1000 mL). Dissolve 38.14 g sodium tetraborate decahydrate ≥99.5% in approximately

800 mL of ultrapure water, cell culture grade, using a magnetic stir plate and bar at high speed. After dissolution of sodium tetraborate, add 6.18 g boric acid ≥99.5% and continue stirring until dissolved. Add 1 N HCl to pH = 8.5 and add water to final volume of 1000 mL. In a sterile laminar flow hood, filter the solution using two 500-mL filter units with 0.22-μm membranes. All subsequent solutions should be prepared in a sterile laminar flow hood and, when indicated, filtered using similar filter units of an appropriate size.

5. Poly-L-lysine solution (1 mg/mL, 100 mL): Dissolve 100 mg of poly-L-lysine hydrobromide, molecular weight 30,000–70,000, in 100 mL of borate buffer and filter. Store at 4 °C when not in use. Poly-L-lysine solution can be reused many times; it should be replaced at the first sign that neurons are not properly adhered to the coverslips.

6. Hank's medium (1×, pH 7.3, 500 mL): Dilute 50 mL of 10× Hank's Balanced Salt Solution (HBSS, no calcium, no magnesium, no phenol red) with 450 mL ultrapure water, cell culture grade, add 10 mL of 1 M HEPES (pH = 7.2–7.5), and filter the solution.

7. Laminin solution (5 μg/mL, 100 mL): Dilute 1 mL of laminin from mouse Engelbreth-Holm-Swarm (EHS) sarcoma (Roche) with 100 mL of Hank's medium and filter the solution. Store at 4 °C when not in use. Laminin can also be reused many times with the same precautions as for poly-L-lysine.

8. Neuronal plating medium (1×, 500 mL): Add 50 mL horse serum (heat inactivated) and 5 mL penicillin–streptomycin (10,000 U/mL) to 450 mL of 1× Dulbecco's Modified Eagle Medium (D-MEM, high glucose, HEPES, no phenol red) and filter. The medium can be stored at 4 °C for 1–2 weeks.

9. Trypsin, 2.5% (10×).

10. Neuronal maintenance medium (50 mL): Prepare fresh medium the day of neuron isolation. Add 1 mL B-27® Supplement (50×), serum free (Gibco™), 500 μL GlutaMAX™ Supplement (100×) (Gibco™), and 500 μL penicillin–streptomycin (10,000 U/mL) to 50 mL of Neurobasal Medium (Gibco™) and filter.

11. Opti-MEM® I Reduced-Serum Medium (1×) (Gibco™).

12. Lipofectamine™ 2000 Transfection Reagent (Invitrogen™).

13. Plasmids encoding RUSH system constructs. Several plasmids for use in the RUSH system are available in Addgene (http://www.addgene.org/) under the list of Franck Perez.

14. Orbital shaker.

15. Autoclave.

16. Horizontal laminar flow hood.

17. Dissecting microscope.

18. Light guide.

19. Dissecting tools (sterilized): microdissecting scissors (flat and curved), fine-tipped forceps (straight and curved).

20. Syringe and needle, 19-gauge $\times$ 1.5″.

21. Vertical laminar flow hood.

22. Water bath at 37 °C.

23. Hemocytometer for counting cells.

24. Tissue culture incubator at 37 °C with humidified, 5 % CO_2 atmosphere.

25. Sterile glass Pasteur pipettes.

26. Sterile filter units, 0.22-μm pore size.

27. Sterile plasticware: 5-mL serological pipettes, 35- and 100-mm dishes, 12-well tissue culture plates, 15- and 50-mL conical centrifuge tubes, 1.5-mL microfuge tubes.

2.2 Endogenous Protein Labeling

1. Anti-Pan-Neurofascin (external) antibody (clone A12/18, purified) (NeuroMab) or any primary antibodies against surface-anchored proteins (external epitope).

2. Mix-n-Stain CF488, CF555, or CF640R antibody labeling kit (Biotium).

3. Refrigerated centrifuge (10,000 $\times g$ at 4 °C).

4. Vortex.

2.3 Live-Cell Imaging to Analyze Sorting of Proteins from Golgi Complex

1. NeutrAvidin Protein (Thermo Scientific™). Prepare a stock of 10 mg/mL of NeutrAvidin using ultrapure water, cell culture grade.

2. D-biotin: prepare a stock of 1-mM D-biotin using ultrapure water, cell culture grade.

3. 35-mm dish type magnetic chamber for 18-mm round coverslips (Quorum Technologies, Inc.).

4. Spinning-disk confocal microscope equipped with 63×, 1.4 NA or 100×, 1.4 NA objectives, EM-CCD camera for digital image acquisition, heating unit and temperature module control.

3 Methods

3.1 RUSH System Construct Design

1. In Addgene, several plasmids deposited by Franck Perez (Curie Institute, Paris, France) for use with the RUSH system are available. A bicistronic expression plasmid (pIRESneo3) was used to generate the expression of two fusion proteins: The ER-hook

(Ii, STIM1-NN, or KDEL retention signal) fused to a core streptavidin, and the reporter fused to SBP.

2. The genes encoding the protein of interest (reporter) must be cloned and fused to SBP. A detailed protocol for subcloning into a RUSH plasmid has been described by Boncompain and Perez [20].

3. The retention and release of the reporter can be followed by live-cell imaging using any of the GFP, mCherry, or TagBFP fluorescent tags available in Addgene.

3.2 Preparation of Coverslips for Neurons

3.2.1 Cleaning and Sterilization of Coverslips

1. At least 1 week prior to isolation of neurons, place 18-mm coverslips in 70 % nitric acid in a ceramic or glass rack inside a chemical fume hood and move gently with forceps to disperse any bubbles.

2. After at least 2 days in nitric acid, remove coverslips to a dish with ultrapure water, cell culture grade. Rinse briefly, change to fresh water, then place dish on an orbital shaker to rinse for 20 min. Rinse five additional times for a total of 2 h.

3. Use a pair of fine forceps to remove each coverslip, touch edges gently to a paper towel to remove excess water, and place in a glass dish.

4. Autoclave dish and maintain in a sterile container or covering.

3.2.2 Coating of Coverslips

1. At least 3 days prior to isolation of neurons, working in a sterile tissue culture hood, place one coverslip into each well of a 12-well plate and add 1 mL per well of poly-L-lysine solution.

2. Leave plates to incubate at 37 °C for at least 2 days. Recover poly-L-lysine, wash each well with at least 0.5 mL ultrapure water, cell culture grade, three times for 10 min per wash, and place in an incubator at 37 °C between washes.

3. Remove all water from last wash and add 1 mL laminin solution to each well. Return plates to 37 °C for 45 min.

4. Recover laminin solution and wash each well with water twice for 15 min per wash at 37 °C.

5. Remove all water from last wash and add 1.2 mL neuronal plating medium to each well.

6. Maintain plates in a CO_2 incubator at 37 °C for up to several days before preparing neurons.

3.3 Isolation of Rat Hippocampal Neurons

The following protocol is a modification of that previously developed by Kaech and Banker [4]. Here we provide a detailed method for dissecting the brains of day-18 embryonic rats and isolating the hippocampus, as well as subsequent steps for dissociating and culturing hippocampal neurons. Coculture with a monolayer of glial feeder cells may be used if desired [4].

Following euthanasia, dissection steps are performed in a sterile horizontal laminar flow cabinet. After hippocampus isolation, perform the remaining steps in a vertical laminar flow hood.

1. Pre-warm an appropriate volume (approximately 15 mL) of neuronal plating medium to 37 °C.

2. On embryonic day 18, sacrifice a pregnant female rat using an approved method of euthanasia and according to an approved animal use protocol. Remove the uterus and transfer to a sterile 100-mm tissue culture dish.

3. Remove each fetus from the uterus and sacrifice by cutting just posterior to the forelimbs using flat scissors, depositing the head and upper torso in a dish of fresh Hank's medium. Repeat for each embryo. Rinse one to two times with fresh Hank's medium to remove any blood or debris.

4. Secure the head using curved forceps and use small curved dissecting scissors to make a lateral incision at the base of the skull, just inferior to the cerebellum. The cerebral hemispheres and cerebellum should be visible beneath the surface (*see* **Note 1**). Make the incision wide enough to allow passage of the brain, then use closed scissors to gently push on the top of the skull in an anterior-to-posterior direction. Allow the brain to slide into a dish of fresh Hank's medium. Repeat for each embryo. Rinse one to two times with fresh Hank's medium and place dish on ice.

5. Transfer a few brains to a sterile glass dish containing fresh Hank's medium and place under a dissecting scope (Fig. 2a). Illumination may be used in order to better distinguish the regions of the brain.

6. Using fine forceps, remove the cerebellum (Fig. 2b) and separate the cerebral hemispheres along the midline (Fig. 2c). For each hemisphere, remove the medial white matter not covered by meninges (Fig. 2c).

7. Carefully remove the meninges (*see* **Note 2**) and locate the hippocampus (Fig. 2d). Look for a thin, curved structure on the medial aspect of the hemisphere that appears slightly more opaque than the surrounding cortex (Fig. 2d). Remove a relatively flat sheet of cortex from the medial face of the hemisphere hemisphere containing the hippocampus along its inferior edge (Fig. 2e).

8. Using a 19-gauge × 1.5″ needle attached to a syringe, carefully cut the hippocampus away from this sheet of cortex (*see* **Note 3** and Fig. 2f). Use one edge of the needle to cut the tissue against the surface of the glass plate.

9. Repeat **steps 5–8** for a few brains at a time, leaving the remaining brains in Hank's medium on ice. Isolated hippocampi may be left in the glass dish until all brains have been dissected.

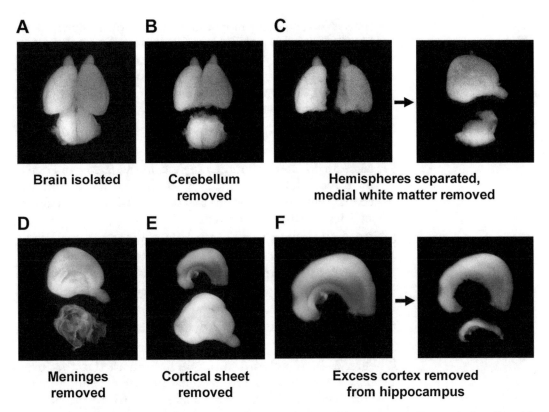

Fig. 2 Hippocampal isolation. (a) Whole brain isolated from a day-18 rat embryo. (b) The cerebellum (*lower*) is removed by making a lateral cut just posterior to the cerebral hemispheres. (c) The hemispheres are separated along the midline, and the medially attached white matter (*lower*) is removed from each. (d) The meninges (*lower*) are peeled from the surface of the hemisphere. (e) A section of cortex (*upper*) containing the hippocampus is removed from the medial face of the hemisphere. The hippocampus can be seen as a more opaque band along the inner edge of this sheet. (f) The hippocampus (*lower*) is cut away from the attached cortical tissue using a needle to complete the isolation

10. Use forceps to transfer the isolated hippocampi to a sterile 35-mm tissue culture dish containing Hank's medium. Use a pipet partially prefilled with fresh Hank's medium to remove the hippocampi from the dish and transfer to a 15-mL conical tube. Add Hank's medium to a volume of 4.5 mL and add 0.5 mL of 2.5 % trypsin. Swirl the tube twice to separate hippocampi. Incubate at 37 °C for a total period of 14–15 min, swirling the tube occasionally.

11. Remove as much trypsin solution as possible without disturbing hippocampi at the bottom of the tube and add 10 mL of fresh Hank's medium. Swirl the tube twice and incubate at room temperature for 6 min, swirling the tube every 2 min.

12. Remove Hank's medium and add 1.5 mL of neuronal plating medium pre-warmed to 37 °C. Use a 5-mL pipet to dissociate hippocampi by pipetting up and down approximately ten

times. Next, pipet the hippocampi up and down approximately ten times using a glass Pasteur pipet whose tip has been narrowed to around 60 % of its original diameter. If chunks are still observed, use a glass Pasteur pipet whose tip has been narrowed to around 30 % and continue pipetting until a homogenous suspension has been achieved.

13. Add 8.5 mL neuronal plating medium for a total volume of 10 mL.

14. Measure cell concentration using a hemocytometer and determine the volume of suspension needed for the desired cell count. Typically, 60,000–80,000 cells per 18-mm coverslip are sufficient for live-cell experiments.

15. Ensure that the cell suspension is uniform by pipetting up and down, then add the appropriate volume to a plate with coated coverslips in neuronal plating medium. Tap the plate on its sides many times to distribute the cells evenly across each well and place in a CO_2 incubator at 37 °C.

16. Pre-warm a sufficient volume of neuronal maintenance medium (at least 1 mL per coverslip) to 37 °C.

17. After 3–4 h, check that most cells have attached to the coverslips, gently remove the neuronal plating medium and any unattached cells using a pipet, and replace with 1 mL per well of neuronal maintenance medium. Neurons may be maintained in this original medium in a CO_2 incubator for several days before transfection.

3.4 Transfection of Rat Hippocampal Neurons (DIV 3-5)

1. Pre-warm appropriate volumes of Neurobasal Medium (3 mL per well to be transfected) and Opti-MEM (200 µL per well) to 37 °C.

2. In a laminar flow hood, prepare DNA mixtures for Lipofectamine 2000-based transfection according to the manufacturer's instructions (100 µL Opti-MEM and approximately 1 µg DNA per well for a 12-well plate) (*see* **Note 4**).

3. Prepare Lipofectamine 2000 mixture using 1.1–1.2 µL Lipofectamine and 100 µL Opti-MEM for each well to be transfected. Incubate for 5 min at room temperature.

4. Combine Lipofectamine and DNA mixtures and incubate for 20 min at room temperature. While incubating, gently recover original culture medium, wash each well twice with 500 µL Neurobasal Medium per well, and add 1 mL of Neurobasal Medium for transfection (*see* **Note 5**).

5. After incubation, add 200 µL of transfection mix to each well and return to a CO_2 incubator at 37 °C for 1 h.

6. During transfection, combine recovered culture medium with an additional ⅓ to ½ volume of fresh neuronal maintenance

medium, filter using a 0.22-μm membrane, and return both filtered medium and Neurobasal Medium to 37 °C.

7. After 1 h of transfection, gently wash each well twice with Neurobasal Medium and add 1 mL of filtered original medium plus fresh neuronal maintenance medium.

8. Add NeutrAvidin (50 μL from a 10 mg/mL stock), a biotin-binding protein, if the RUSH system plasmids have been transfected (*see* **Note 6**).

9. Maintain plates in a CO_2 incubator until ready for use. It is not necessary to change the culture medium during this time. Typical transfection efficiency is approximately 5%.

3.5 Staining of the AIS for Live-Cell Imaging

This protocol may also be used for axonal and dendritic surface staining of proteins for live-cell imaging through selection of an antibody to the extracellular portion of a cell surface protein. Follow the manufacturer's protocol for covalent coupling of each antibody to the Mix-n-Stain CF series dyes.

1. Centrifuge 90 μL of Anti-Pan-Neurofascin (external) antibody (clone A12/18, purified) solution at $10,000 \times g$ for 30 min.

2. After centrifugation of the antibody, perform all subsequent steps in a laminar flow hood.

3. Take the supernatant and incubate with the 10× Mix-n-Stain (10 μL) solution, pipetting up and down a few times.

4. Transfer the entire solution to the vial containing the CF555, CF488, or CF640R dyes and vortex the vial for a few seconds.

5. Incubate the mix for 30 min at room temperature in a dark room.

6. Prepare 10-μL aliquots of the mixture and store at −20 °C.

7. Prepare a 1:50 dilution of Anti-Pan-Neurofascin-CF with neuronal maintenance medium and pre-warm the solution to 37 °C. One aliquot is enough for one coverslip in a 12-well plate (10 μL of the Neurofacin-CF640R plus 500 μL of neuronal maintenance medium).

8. Remove the medium from one coverslip with neurons on DIV 6-14 and add the 500 μL of antibody solution. Return the plate with neurons to a CO_2 incubator at 37 °C for 30 min to 1 h (*see* **Note 7**).

9. Recover the 500 μL of antibody solution and transfer to another coverslip for the next experiment. The antibody solution can be used many times the same day. Wash the coverslip twice with pre-warmed neuronal maintenance medium and add 1 mL of the same medium to the coverslip.

10. Maintain the plate in the CO_2 incubator at 37 °C until preparing the coverslips for live-cell imaging.

**3.6 Live-Cell
Imaging Using
the RUSH System**

1. Turn on the spinning-disk confocal microscope, set the temperature to 37 °C using a temperature controller connected to a heating device, and wait until the temperature is stabilized to 37 °C. Pre-warm the oil used for objectives to 37 °C.

2. In a vertical laminar flow hood, remove the coverslip from the plate using sterile fine forceps, place the coverslip in a 35-mm dish type magnetic chamber for incubating 18-mm round coverslips during live-cell imaging, and add 500 μL of the original medium.

3. Clean the bottom surface of the coverslip in the magnetic chamber with 70 % ethanol before imaging.

4. In the spinning-disk microscope, add a pre-warmed drop of oil to a 63× oil or 100× oil objective and place the magnetic chamber in the microscope.

5. Focus and find neurons expressing constructs of interest. Choose optimal conditions of laser power and exposure for imaging those particular constructs.

6. Neurites can be identified by expression of GFP, RFP, CFP, or similar soluble fluorescent proteins, tagged tubulin, or tagged plasma membrane proteins in the transfection mix, and/or endogenous plasma membrane proteins stained with the CF-series dyes (Figs. 3 and 4).

7. Find transfected neurons expressing proteins of interest using the RUSH system (*see* **Note 8**).

8. Add 40 μM D-biotin to neurons. Dilute 24 μL of D-biotin (1-mM stock) with neuronal maintenance medium (100 μL) and add to the chamber containing 500 μL of medium. If necessary, adjust the focus again in the microscope.

9. If the cargo molecules have been retained by interaction with an ER resident protein, it will take around 10–15 min to release the cargo molecules from the ER, and strong staining of the Golgi complex will appear (Fig. 4a–c).

10. After accumulation of the cargo molecules at the Golgi complex, start the live-cell imaging to analyze trafficking of transmembrane proteins to the axon and dendrites (Fig. 4a).

**3.7 Analysis
of Live-Cell Imaging
Data**

Trafficking of proteins to the axon and dendrites can be studied using kymograph analysis.

1. Open images containing the frames in ImageJ (Wayne Rasband, NIH; http://imagej.nih.gov).

2. Isolate axon and dendrites from a neuron. Select a line width between 20 and 40 pixels (Edit/Options/Line width). Then use the segmented line tool to select a region from the axon or dendrite (Fig. 5a, b). Straighten the selected line (Edit/Selection/Straighten) and select "Process Entire Stack" (Fig. 5c). Draw each

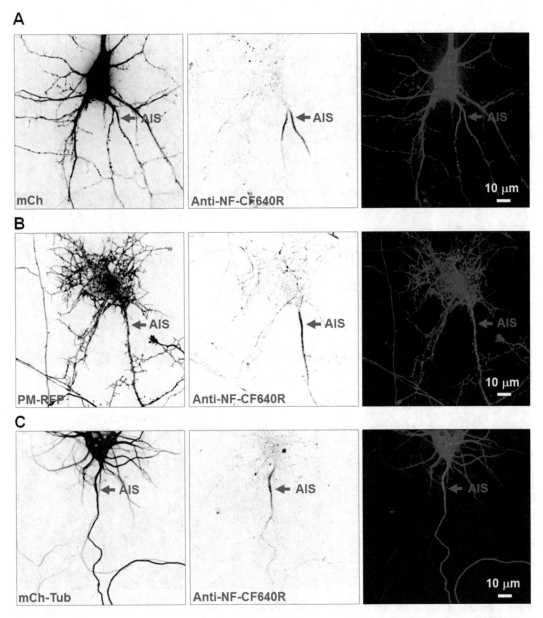

Fig. 3 Expression of neuronal markers and staining of endogenous proteins. (**a–c**) One-frame images corresponding to live neurons transfected on DIV 4 with plasmids encoding mCherry (mCh) (**a**), 13 amino acids of the N-terminal sequence of the myristoylated and palmitoylated Lyn kinase to target mRFP to the plasma membrane (PM-RFP) (**b**), and mCherry-tagged alpha-tubulin (mCh-Tub) (**c**), and stained for the AIS with CF640R-conjugated antibody to neurofascin (Anti-NF-CF640R) on DIV 7 (**a–c**). Transgenic proteins are shown in negative grayscale (*left panels*) and *red* in the merged images (*right panels*), and surface neurofascin in negative grayscale (*middle panels*) and *blue* in the merged images (*right panels*). *Blue arrows* point to the AIS. Scale bar, 10 µm. Notice the staining of the whole neuronal cytoplasm for mCh and mCh-Tub, and the whole plasma membrane for PM-RFP. Staining with anti-NF-CF640R, on the other hand, is restricted to the AIS

A

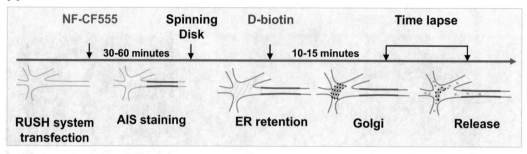

B

+ D-biotin 15 minutes

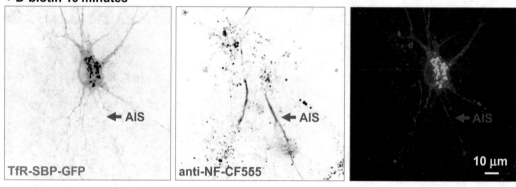

C

+ D-biotin 20 minutes

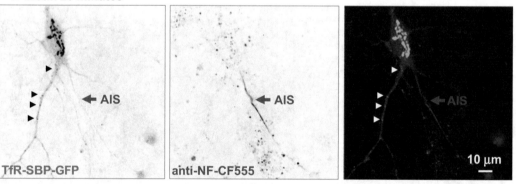

Fig. 4 Expression of RUSH system constructs to analyze sorting of proteins from the Golgi complex. (**a**) Schematic representation of the procedure to analyze protein sorting from the Golgi complex to dendrites and axon using the RUSH system. (**b**, **c**) Neurons transfected on DIV 4 with a bicistronic expression plasmid encoding two fusion proteins: the KDEL ER-retention signal fused to a core streptavidin ("hook") and the transferrin receptor (TfR) fused to SBP and GFP (TfR-SBP-GFP). NeutrAvidin was added to the culture medium after transfection to compete with the D-biotin present in the neuronal maintenance medium. Neurons on DIV 7 were surface-stained by incubation with CF555-conjugated antibody to neurofascin (NF-CF555) for 45 min at 37 °C and analyzed using a spinning-disk confocal microscope. Images show the expression of TfR-SBP-GFP (negative grayscale in *left* and *middle panels*, and *green* in *right panels*) after 15 min (**b**) and 20 min (**c**) of incubation with D-biotin. NF-CF555 staining is shown in grayscale (*middle panels*) and *red* in the merged images (*right panels*). *Red arrows* point to the AIS. Scale bar 10 μm. Notice the accumulation of TfR-SBP-GFP at the Golgi complex after 15 min of D-biotin incubation. After 20 min of D-biotin incubation, some carriers containing TfR-SBP-GFP can be observed in dendrites (*arrowheads*), consistent with the somatodendritic localization of the TfR. A similar approach can also be used for studying axonal proteins sorting from the Golgi complex

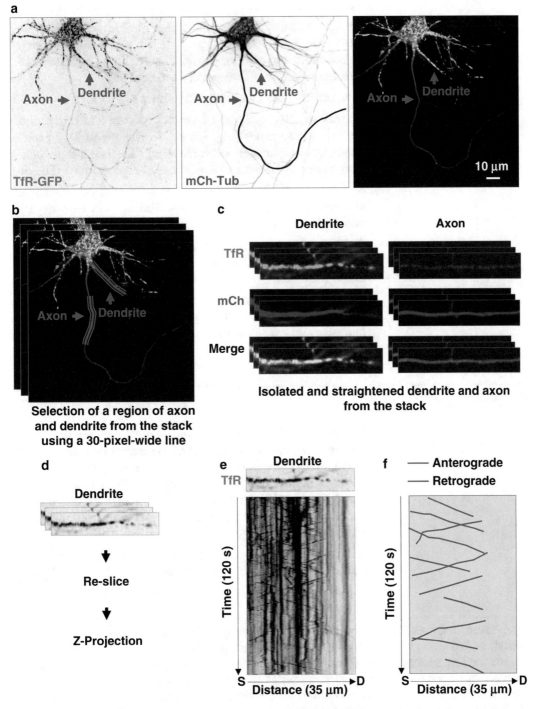

Fig. 5 Kymograph analysis of the transport of proteins to dendrites and axon. (**a**) Single-frame images corresponding to a live neuron on DIV 7 co-transfected on DIV 4 with GFP-tagged transferrin receptor (TfR-GFP) (negative grayscale in *left panel*, *green* in merged image in *right panel*) and mCherry-tubulin (mCh-Tub) (negative grayscale in *middle panel*, *red* in merged image in *right panel*). Note the somatodendritic distribution of TfR. Live-cell imaging was performed using a spinning-disk confocal microscope (Intelligent Imaging Innovations). Digital images were acquired with an Evolve electron-multiplying charge-coupled device (EM-CCD)

line beginning in the soma and ending in the axon or dendrite in order to visualize anterograde transport from left to right and retrograde transport from right to left. Use the same parameters of line width and axon or dendrite length for all experiments for direct comparison of transport parameters.

3. Save isolated axon and dendrite frames as a TIFF file.

4. To quantify the transport of cargoes to axon and or dendrites, generate kymographs (Fig. 5d, e). Re-slice the isolated axon or dendrite stack (Image/Stacks/Reslice). Generate a Z-projection from the re-sliced stack (Image/Stacks/Z-projection) (Fig. 5e).

5. Stationary particles are observed as vertical lines; anterograde and retrograde transport are observed as lines with negative and positive slopes, respectively (Fig. 5e).

6. Anterograde and retrograde transport can be manually represented by removing the stationary particles from the kymograph (Fig. 5f).

7. The number of stationary, anterograde, and retrograde particles can be determined through the quantification of the number of events from kymographs of the same duration and neurite length.

4   Notes

1. If the cerebral hemispheres and cerebellum are not visible through the skull, or if the base of the skull is rigid and difficult to cut using scissors, the embryos obtained may be older than 18 days. Brains obtained from embryos obtained after day 18 may be more difficult to dissect, and it may be necessary to make an additional incision in the skull along the midline in order to remove the brain.

2. It is typically easiest to peel away the meninges by creating a small tear at either the anterior or posterior end of the hemisphere,

Fig. 5 (continued) camera (Photometrics). Images of TfR-GFP and mCh-Tub were sequentially acquired at 500-ms intervals for 120 s. *Arrows* point to the axon. Scale bar 10 μm. (**b–f**) An example of selection and isolation of dendrite and axon segments for kymograph analysis. (**b**) Lines 30 pixels wide and 35 μm long in a dendrite (*magenta region*) and axon (*cyan region*) were selected with orientation S→D and S→A from a 240-frame dual-color image. (**c**) Selected dendrite and axon lines from the 240-frame dual-color image were straightened. TfR-GFP and mCh-Tub images are shown separately and merged. (**d**) The straightened dendrite was re-sliced to produce a Z-projection in order to generate a kymograph. (**e**) Kymograph generated from a dendrite, where the X-axis is distance (35 μm) and Y-axis is time (120 s). *Vertical lines* correspond to stationary particles that did not move during the entire period of imaging. Lines with negative slopes and positive slopes correspond to anterograde and retrograde movement, respectively, of TfR-GFP in the dendrite during the 120 s of recording. (**f**) Representation of anterograde (*magenta lines*) and retrograde (*blue lines*) particles from the kymograph

then using one pair of forceps to pull on the membranes while gently holding the hemisphere stationary with the other. If removed carefully, the meninges will peel away from the cortex as a single sheet.

3. The hippocampus is most easily identified and separated when what was formerly the medial surface of the hemisphere is facing upward. There are various techniques for removing the hippocampus using the needle. Generally, it is easiest to use a pair of forceps in the nondominant hand to hold the cortical sheet stationary while the hippocampus is cut away. The forceps can either be placed directly on the cortex or in the growing incision between the hippocampus and remaining cortical tissue.

4. If transfecting more than two 12-well plates, it may be necessary to wash each plate once prior to the 5-min Lipofectamine incubation step. After washing once, return plates to incubator and resume washing during the 20-min incubation of the Lipofectamine–DNA mixture.

5. Primary neurons are fragile and susceptible to detachment of neurites if media are added too forcefully. Examine neurons after initial washes and again after transfection to determine whether cell survival is affected by either step. If cell density decreases noticeably after washing, be sure to wash gently by touching the pipet tip to the side of the well and adding the medium slowly.

6. B27 supplement contains D-biotin, which can release the cargo molecules from the donor compartment when using the RUSH system. This dissociation can be prevented by addition of NeutrAvidin to the medium on the day of transfection.

7. AIS staining in young neurons (DIV 4–6) will be weak compared to old neurons (DIV 7–21) because Neurofascin accumulates during neuronal development. If staining on DIV 6 is desired, perform a longer incubation with the anti-Neurofascin-CF-series dyes (1 h). If staining for Neurofascin is performed on DIV 10 or later, 30 min of incubation will be sufficient to visualize the AIS.

8. In transfected neurons, expression of a protein of interest using the RUSH system will appear weak if the protein has been "hooked" to an ER resident protein, but after adding D-biotin for 10–15 min, the protein will accumulate in the Golgi complex (Fig. 4b, c).

Acknowledgement

This work was funded by the Intramural Program of NICHD, NIH (ZIA HD001607).

References

1. Lasiecka ZM, Yap CC, Vakulenko M, Winckler B (2009) Compartmentalizing the neuronal plasma membrane: from axon initial segments to synapses. Int Rev Cel Mol Biol 272: 303–389

2. Tahirovic S, Bradke F (2009) Neuronal polarity. Cold Spring Harb Perspect Biol 1:a001644

3. Rasband MN (2010) The axon initial segment and the maintenance of neuronal polarity. Nat Rev Neurosci 11:552–562

4. Kaech S, Banker G (2006) Culturing hippocampal neurons. Nat Protoc 1:2406–2415

5. Lasiecka ZM, Winckler B (2011) Mechanisms of polarized membrane trafficking in neurons – focusing in on endosomes. Mol Cell Neurosci 48:278–287

6. Bonifacino JS (2014) Adaptor proteins involved in polarized sorting. J Cell Biol 204:7–17

7. Hanus C, Ehlers MD (2008) Secretory outposts for the local processing of membrane cargo in neuronal dendrites. Traffic 9:1437–1445

8. Nixon RA, Cataldo AM (1995) The endosomal-lysosomal system of neurons: new roles. Trends Neurosci 18:489–496

9. Golgi C (1898) Intorno alla struttura delle cellule nervose. Bollettino della Società Medico-Chirurgica di Pavia 13:3–16

10. Ashby MC, Ibaraki K, Henley JM (2004) It's green outside: tracking cell surface proteins with pH-sensitive GFP. Trends Neurosci 27: 257–261

11. González-González IM, Jaskolski F, Goldberg Y, Ashby MC, Henley JM (2012) Measuring membrane protein dynamics in neurons using fluorescence recovery after photobleach. Methods Enzymol 504:127–146

12. Hildick KL, González-González IM, Jaskolski F, Henley JM (2012) Lateral diffusion and exocytosis of membrane proteins in cultured neurons assessed using fluorescence recovery and fluorescence-loss photobleaching. J Vis Exp 60:e3747

13. Matlin KS, Simons K (1983) Reduced temperature prevents transfer of a membrane glycoprotein to the cell surface but does not prevent terminal glycosylation. Cell 34:233–243

14. Saraste J, Kuismanen E (1984) Pre- and post-Golgi vacuoles operate in the transport of Semliki Forest virus membrane. Cell 38:535–549

15. Bergmann JE (1989) Using temperature-sensitive mutants of VSV to study membrane protein biogenesis. Methods Cell Biol 32:85–110

16. Presley JF, Cole NB, Schroer TA, Hirschberg K, Zaal KJ, Lippincott-Schwartz J (1997) ER-to-Golgi transport visualized in living cells. Nature 389:81–85

17. Hirschberg K, Miller CM, Ellenberg J, Presley JF, Siggia ED, Phair RD, Lippincott-Schwartz J (1998) Kinetic analysis of secretory protein traffic and characterization of Golgi to plasma membrane transport intermediates in living cells. J Cell Biol 143:1485–1503

18. Rivera VM, Wang X, Wardwell S, Courage NL, Volchuk A, Keenan T, Holt DA, Gilman M, Orci L, Cerasoli F Jr, Rothman JE, Clackson T (2000) Regulation of protein secretion through controlled aggregation in the endoplasmic reticulum. Science 287:826–830

19. Boncompain G, Divoux S, Gareil N, de Forges H, Lescure A, Latreche L, Mercanti V, Jollivet F, Raposo G, Perez F (2012) Synchronization of secretory protein traffic in populations of cells. Nat Methods 9:493–498

20. Boncompain G, Perez F (2014) Synchronization of secretory cargos trafficking in populations of cells. Methods Mol Biol 1174:211–223

Chapter 3

Imaging Golgi Outposts in Fixed and Living Neurons

Mariano Bisbal, Gonzalo Quassollo, and Alfredo Caceres

Abstract

Here we describe the use of confocal microscopy in combination with antibodies specific to Golgi proteins to visualize dendritic Golgi outposts (GOPs) in cultured hippocampal pyramidal neurons. We also describe the use of spinning disk confocal microscopy, in combination with ectopically expressed glycosyltransferases fused to GFP variants, to visualize GOPs in living neurons.

Key words Neurons, Cultures, Dendrites, Golgi outposts, Confocal microscopy, Live imaging

1 Introduction

Cultures of embryonic rat hippocampal pyramidal neurons [1] have become a widely used model for analyzing the mechanisms underlying the development of axons and dendrites. Using this cell system it has been established that membrane trafficking is central for neuronal polarization [2, 3]. In fact, elements of the secretory pathway, such as the Golgi apparatus (GA) have a pivotal role in the establishment and maintenance of neuronal polarity and synaptic plasticity [4, 5]. In neurons, the GA not only consists of perinuclear cisternae but also of satellite tubule-vesicular structures designated as Golgi outposts (GOPs) that localize to dendrites [6–8]. GOPs have been implicated in dendritic morphogenesis and as stations for the local delivery of post-synaptic membrane receptors [5]. Here, we describe methods for the visualization of GOPs in fixed and living cultured hippocampal pyramidal neurons.

2 Materials

2.1 Neuron Isolation, Growth, and Plating

1. CMF-HBSS (Calcium-, magnesium-, and bicarbonate-free Hank's balanced salt solution (BSS) buffered with 10 mM HEPES, pH 7.3).

William J. Brown (ed.), *The Golgi Complex: Methods and Protocols*, Methods in Molecular Biology, vol. 1496,
DOI 10.1007/978-1-4939-6463-5_3, © Springer Science+Business Media New York 2016

2. Neuronal Plating Medium (MEM supplemented with glucose (0.6 % wt/vol) and containing 10 % (vol/vol)) horse serum.

3. Neurobasal/B27 Medium: Prepare according to the manufacturer's instructions by supplementing Neurobasal Medium (with or without Phenol red) (Gibco) with GlutaMAX-I CTS supplement (Gibco), B27 serum-free supplement (Gibco) and penicillin–streptomycin.

4. 2.5 % (wt/vol) trypsin; aliquot for single use and store at –20 °C.

5. Poly-L-lysine (1 mg/ml) in 0.1 M borate buffer pH 8.5 and filter-sterilize (prepared immediately before use): Poly-L-Lysine, molecular weights 30,000–70,000 kDa; 0.1 M Borate buffer, pH 8.5 (prepared from boric acid and sodium tetraborate).

6. 5 mM AraC (Cytosine-1-β-D-arabinofuranoside); aliquot and store at 4 °C.

7. Lipofectamine 2000.

8. Coverslips Assistant #1 de 25 mm (Carolina Biological Supply Company).

9. 70 % nitric acid (wt/wt).

10. Fire-polished glass Pasteur pipette.

2.2 Expression Plasmids

1. For neuronal live imaging [9] of the Golgi apparatus and dendritic GOPs we use several cDNA plasmid coding for the N-terminal domains (cytosolic tail, transmembrane domain, and a few amino acids of the stem region) of Golgi resident enzymes (e.g., glycosyltransferases; [10]) fused to the N terminus of the enhanced yellow fluorescent protein (YFP) or the red fluorescent protein, mCherry. The following constructs are routinely used for labeling the GA and GOPs in living or fixed cultured neurons: (1) Sialyl-transferase 2 (SialT2), a marker of the cis-Golgi; (2) Galactosyl-transferase 2 (GalT2), a marker of the media- and trans-GA; and (3) β-1, 4-N-acetylgalactosaminyltransferase (GalNAcT), a marker of the trans-GA. For further details [8, 10–12].

2.3 Antibodies

1. A mouse monoclonal antibody (mAb) against GM130, a marker of the cis-Golgi compartment (Clone 35/GM130; BD Biosciences) diluted 1:250.

2. A mAb against Mannosidase II, a marker of the Golgi apparatus (clone 53FC3, ab24565; AbCam), diluted 1:100.

3. A rabbit polyclonal antibody against TGN38, a marker of the TGN (Product Number T9826, Sigma Chemical Co.) diluted 1:250.

4. A mAb against MAP2 (clone AP20, M 1406, Sigma Chemical Co.) diluted 1:500.

5. Goat anti-mouse IgG-Alexa 488 conjugated (A-11001, Life Technologies).

6. Goat anti-rabbit IgG-Alexa 546 conjugated (A-11035, Life Technologies, Thermo Fisher Scientific Inc. Rockford, IL, USA).

7. Goat anti-mouse IgG-Alexa 633 conjugated (A-21050, Life Technologies, Thermo Fisher Scientific Inc. Rockford, IL, USA).

8. Goat anti-rabbit IgG-Alexa 633 conjugated (A-21070, Life Technologies, Thermo Fisher Scientific Inc. Rockford, IL, USA).

2.4 Immunofluorescence

1. Phosphate buffered saline (PBS), pH 7.2–7.4.

2. 4 % paraformaldehyde (PFA), 4 % sucrose in PBS buffer.

3. 0.2 % Triton X-100 in PBS.

4. 5 % bovine serum albumin (BSA) in PBS.

5. 1 % BSA in PBS.

6. FluorSave™ Reagent (or similar medium) for mounting coverslips on microscope slides.

2.5 Microscopy

1. Conventional confocal microscopes: Zeiss Pascal or Olympus FV300.

2. Spectral confocal microscope: Olympus FV1000.

3. Spinning disk confocal microscope: Olympus IX81 plus Disk Spinning Unit (DSU).

3 Methods

3.1 Preparation of Coverslips

Cultured hippocampal neurons are grown attached to glass coverslips (12, 15, 25 mm in diameter) coated with different substrates (e.g., poly-lysine, laminin, and tenascin). Careful preparation of glass coverslips that serve as substrates for neuronal growth is a key step to assure appropriate development of axons and dendritic arbors. For most experiments we coated glass coverslips with poly-L-lysine (molecular weight 30,000–70,000).

1. Place 25-mm glass coverslips in porcelain racks in concentrated 70 % nitric acid for 24 h (range 18–36 h).

2. Wash the coverslips (still in the racks) with distilled water three times/30 min each and then with Milli Q (type1 ultrapure) water two more times for 30 min with constant stirring.

3. Place the racks in a glass beaker covered with aluminum foil and sterilize in an oven at 220 °C for 4–6 h.
   Carry out all the following procedures sterile under a laminar flow hood.

4. Place coverslips in a 35-mm cell-culture dish or in a 6-multi well culture plate.

5. Cover each coverslip with the Poly-L-lysine solution (minimum 150 μl per coverslip). Leave at room temperature in the laminar flow hood (~20–25 °C) overnight.

6. Rinse the coated coverslips with sterile ultrapure MilliQ water every 20 min over a period of 2 h.

7. Discard water from the last wash and add Neuronal Plating Medium. Place the culture dishes in a CO_2 incubator at least 2 h before plating neurons. (Dishes can be stored in the incubator for several days before using them.)

3.2 Primary Cultures of Rat Hippocampal Neurons

1. Extract the brain from embryonic day 18 (E18) rat embryos and kept them in a dish with ice-cold Hank's.

2. Under a dissecting microscope, carefully remove the meninges from cerebral hemispheres and then dissect out the hippocampus. Collect the hippocampi in a dish containing ice-cold CMF-HBSS.

3. Place the extracted hippocampi in a 15 ml centrifugal tube and load up to 2.7 ml of CMF-HBSS. Add 0.3 ml of 2.5 % trypsin and incubate for 15 min in a water bath at 37 °C.

4. Remove the trypsin solution, and rinse the hippocampi two times with 5 ml of CMF-HBSS.

5. Using a fire-polished glass Pasteur pipette, mechanically dissociate hippocampi pipetting them up and down (five to ten times). Then repeat this procedure with narrowed flame-polished pipette. At this point, there should be no chunks or tissue left.

6. Plate the desired number of cells (between 100,000 and 150,000) in the culture dishes containing the coated coverslips in the Neuronal Plating Medium and place them in the incubator at 37 °C with 5 % CO_2 for 2–4 h.

7. Replace the Neuronal Plating Medium for warmed (37 °C) Neurobasal/B27 Medium and incubate at 37 °C with 5 % CO_2.

8. After plating for 72 h, add AraC (Cytosine-1-β-D-arabinofuranoside) to a final concentration of 5 μM to inhibit proliferation of glial cells. To maintain the culture, every 3 days replace 1/3 medium with fresh Neurobasal/B27 Medium. For further details see ref. [1] (see **Note 1**).

3.3 Transient Transfection of Cultured Hippocampal Neurons

For visualization of GOPs in living and fixed neurons we transiently transfect 14 days in vitro (DIV) cultured hippocampal neurons (see **Note 2**) with cDNAs encoding one of several resident Golgi glycosyltransferases (see above) (see **Note 3**).

1. Under a laminar flow hood, collect and store the Neurobasal/ B27 medium from the cultured neurons and replace it with warm 2 ml Neurobasal.

2. For one (1) 35-mm cell-culture dish:

   Add to one tube 62.5 μl Neurobasal + 2 μl Lipofectamine 2000.

   Add to another tube 62.5 μl Neurobasal + 0.5–1 μg DNA.

3. Incubate for 5 min at room temperature (RT).

4. Combine the contents of the tubes from **steps 2**- with gentle mixing.

5. Incubate for 20–25 min at RT.

6. Add the transfection mix to the cells and swirl gently.

7. Incubate for 1–2 h at 37 °C with 5 % CO_2.

8. Replace the transfection media with the original Neurobasal/B27 medium and return cells to the incubator.

3.4 Immunofluorescence of Fixed Cells

GOPs are also visualized in fixed cultures with specific antibodies to Golgi proteins. All steps for immunofluorescence labeling are performed at room temperature. For immunofluorescence of fixed cells, cultures are usually grown on 12-mm glass coverslips

1. *Fixation.* Rinse the cells with PBS at 37 °C. Remove PBS and immediately fix using 4 % PFA (paraformaldehyde)—4 % sucrose in PBS buffer at 37 °C, for 20 min. Paraformaldehyde is a popular fixative and will usually result in better preservation of cellular morphology than methanol or acetone.

2. Wash cells thoroughly (3–5 washes × 5 min in PBS).

3. *Permeabilize* with 0.2 % Triton TX-100 in PBS for 10 min.

4. Wash cells thoroughly (3–5 washes × 5 min in PBS).

5. *Block* by incubation with 5 % BSA (bovine serum albumin) in PBS for 1 h.

6. *Incubate with primary antibody.* Prepare primary antibody in 1% BSA in PBS. Incubate for 1 h at room temperature or overnight at 4 °C.

7. Wash cells thoroughly (3–5 washes × 5 min in PBS).

8. *Incubate with secondary antibody.* Prepare secondary antibody (dilution 1:200) solution as in the case of primary antibodies. Incubate 1 h at room temperature. Longer incubation may increase background.

9. Wash cells thoroughly (3–5 washes × 5 min in PBS).

10. *Mounting.* We routinely use FluorSave™ Reagent.

3.5 Microscopy

3.5.1 Imaging Fixed Samples

To visualize fixed stained cells we acquired high resolution images (1024 × 1024 image size, 12-bit per pixel) using either a conventional (Zeiss Pascal) or a spectral (Olympus Fluoview 1000) inverted confocal microscope with an oil immersion PlanApo

60×/1.4 objective. We collect a 5–10 μm-deep Z stack along the optical axis to obtain voxel sizes of approximately $0.2 \times 0.2 \times 0.2$ μm. The image files were processed and analyzed with ImageJ software (*see* **Note 4**).

3.5.2 Live Cell Imaging

General imaging considerations: For time-lapse fluorescence microscopy, we used an Olympus IX81 inverted microscope equipped with a Disk Spinning Unit (DSU) and epi-fluorescence illumination (150W Xenon Lamp). We typically obtain 3–5 optical sections (700 nm each) every 2 s during a time period of 3–5 min. However, since dendrites are sometimes quite thin (1.0 μm thickness), the GA and GOPs can be completely captured and imaged with a single optical slice. In these cases, time-lapse sequences are acquired at a continuous rate of 2–5 frames per second during 3–5 min.

Objective

Cells plated on coverslips are well within the working distance of all microscope objectives, including high-NA PlanApo objectives, which offer the greatest optical correction and superior light gathering ability. These high-NA objectives are ideal for live cell fluorescence imaging. For instance, in our set up we use a 60× PlanApo 1.4 NA. In some cases, we have used TIRFM objectives with higher NA (1.45).

Cameras

Charge-coupled device (CCD) cameras have become the standard detectors for live cell imaging. CCD cameras provide high sensitivity and linear response over their dynamic range. Our DSU microscope set up is coupled with an ANDOR iXon3 CCD camera. We have also imaged GOPs using other cameras, such as the ORCA-AG (Hamamatsu) or the ORCA-ER (Hamamatsu).

Environmental Control

For recordings we use open chambers (Fig. 1), which allow cells to be maintained in bicarbonate-based media, inside a stage top incubator (INU series, TOKAI HIT) with 37 °C controlled temperature, 5% CO_2 and humidity (enriched atmosphere). In the chambers, the coverslips (25 mm diameter) with the attached cells rests face-up on a base plate. A rubber ring above the coverslip forms a well to hold the culture medium (Fig. 1). We have also used other devices such as a Harvard micro-incubator (Harvard Instruments, model PDMI-2). To avoid focus drift due to changes in objective temperature when contacting the sample we routinely use a lens heater collar.

Recording Media

We use medium without phenol red to avoid toxic breakdown products when exposed to light. For this, the cells are cultured in phenol red free-medium; alternatively, the cells can be grown in phenol red-free medium.

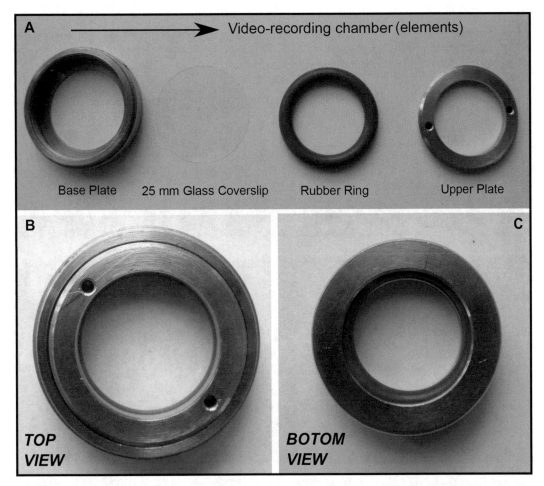

Fig. 1 (**a**) Components of the open chambers used for video recordings. They include: (a) base plate; (b) 25 mm coverslips, where the cells are grown; (c) a rubber ring, and (d) the upper plate. (**b**) Top view of the assembled chamber. (**c**) Bottom view of the assembled chamber

4 Notes

1. A major problem for the appropriate detection and visualization of GOPs is the quality of the neuronal culture. In our hands the best protocol for culturing embryonic hippocampal pyramidal neurons is the one developed and used in Gary Banker's lab [1].

2. Another important aspect for detection and visualization of GOPs is the age of the culture. GOPs are abundant in neurons with well-developed dendrites (long and highly branched MAP2 + neurites). These dendrites are found in neurons that have developed in culture for at least 10 DIV. We routinely use

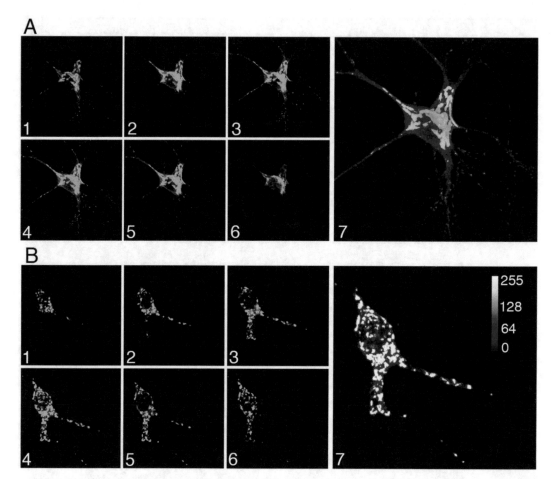

Fig. 2 (**a**) Serial (1–6) confocal micrographs showing the morphology of the GA and GOPs in a 14 DIV hippo-campal pyramidal neuron transfected with SialT2-YFP (7). Pseudocolor (Fire LUT) maximal projection image obtained from the serial confocal micrographs shown previously. Note the presence of GA-derived tubules (*arrowheads*) (*large arrows*) and of small SialT2 + vesicular structures (GOPs, *arrows*) localized along den-drites. The neuron shows no signs of somatic GA fragmentation. (**b**) Another example of a 14 DIV neuron expressing SialT2-YFP. The Fire LUT (7) clearly reveals that this cell express much higher levels of ectopic protein than the one shown in panel (**a**). Neurons expressing high levels of ectopic resident Golgi glycosyltrans-ferases quite frequently display somatic GA fragmentation and dispersal into dendrites. We routinely do not use neurons with fragmented GA

14-21 DIV neurons. GOPS are difficult to detect in young cultures (less than a week).

3. GOPS are easily visualized by ectopic expression of GA resi-dent proteins, such as glycosyltransferases. However, when using this procedure care should be taken to avoid excessive expression of the ectopic protein. High expression levels of Golgi-resident enzymes can produce Golgi fragmentation and dispersal into dendrites (Fig. 2). Because of this, we only use neurons with moderate levels of expression and without signs of somatic Golgi fragmentation.

4. To highlight detection of small GOPs (less than 1 μm), either labeled by ectopic expression of fusion proteins or specific antibodies, we routinely used a look-up-table (LUT) called FIRE (scale thermal) available within the LUT table of FIJI ([13] *see* also Fig. 2).

Acknowledgements

Research at the Laboratory of Neurobiology at INIMEC-CONICET-UNC is supported by grants from FONCyT, CONICET to A.C. M.B. is supported by a return home grant from ISN and FONCyT grant (PICT D 2013-1525). A.C. and M.B. are staff scientists from CONICET.

References

1. Kaech S, Banker G (2006) Culturing hippocampal neurons. Nat Protoc 1:2406–2415

2. Conde C, Cáceres A (2009) Microtubule assembly, organization and dynamics in axons and dendrites. Nat Rev Neurosci 10:319–332

3. Namba T, Funayashi Y, Nakamuta S, Xu C, Takano T, Kaibuchi K (2015) Extracellular and intracellular signalling for neuronal polarity. Physiol Rev 95:995–1024

4. Horton AC, Ehlers M (2003) Neuronal polarity and trafficking. Neuron 40:277–295

5. Ehlers MD (2013) Dendritic trafficking for neuronal growth and plasticity. Biochem Soc Trans 41:1365–1382

6. Horton AC, Rácz B, Monson EE, Lin AL, Weinberg RJ, Ehlers MD (2005) Polarized secretory trafficking directs cargo for asymmetric dendrite growth and morphogenesis. Neuron 48:751–771

7. Ye B, Zhang Y, Song W, Younger SH, Jan LY, Jan YN (2007) Growing dendrites and axons differ in their reliance on the secretory pathway. Cell 130:717–729

8. Quassollo G, Wojnacki J, Salas DA, Gastaldi L, Marzolo MP, Conde C, Bisbal M, Couve A, Cáceres A (2015) A RhoA signaling pathway regulates dendritic Golgi outpost formation. Curr Biol 25:971–982

9. Kaech S, Huang CF, Banker G (2012) General considerations for live imaging of developing hippocampal neurons in culture. Cold Spring Harb Protoc 201:312–318

10. Maccioni HFC, Quiroga R, Spessott W (2011) Organization of the synthesis of glycolipid oligosaccharides in the Golgi complex. FEBS Lett 585:1691–1698

11. Giraudo CG, Rosales Fritz VM, Maccioni HF (1999) Ga2/Gm2/Gd2 synthase localizes to the trans-Golgi network of cho-k1 cells. Biochem J 342:633–640

12. Giraudo C, Daniotti J, Maccioni H (2001) Physical and functional association of glycolipid N-acetyl-galactosaminyl and galactosyl transferases in the Golgi apparatus. Proc Natl Acad Sci U S A 98:1625–1630

13. Eliceiri KW, Berthold MR, Goldberg IG, Ibáñez L, Manjunath BS, Martone ME, Murphy RF, Peng H, Plant AL, Roysam B, Stuurman N, Swedlow JR, Tomancak P, Carpenter AE (2012) Biological imaging software tools. Nat Methods 9:697–710

Chapter 4

Analysis of Arf1 GTPase-Dependent Membrane Binding and Remodeling Using the Exomer Secretory Vesicle Cargo Adaptor

Jon E. Paczkowski and J. Christopher Fromme

Abstract

Protein–protein and protein–membrane interactions play a critical role in shaping biological membranes through direct physical contact with the membrane surface. This is particularly evident in many steps of membrane trafficking, in which proteins deform the membrane and induce fission to form transport carriers. The small GTPase Arf1 and related proteins have the ability to remodel membranes by insertion of an amphipathic helix into the membrane. Arf1 and the exomer cargo adaptor coordinate cargo sorting into subset of secretory vesicle carriers in the model organism *Saccharomyces cerevisiae*. Here, we detail the assays we used to explore the cooperative action of Arf1 and exomer to bind and remodel membranes. We expect these methods are broadly applicable to other small GTPase/effector systems where investigation of membrane binding and remodeling is of interest.

Key words GTPase, Arf1, Membrane remodeling, Membrane binding, Membrane trafficking, Membrane fission, Membrane scission, Cargo adaptor, Coat protein

1 Introduction

Protein–membrane and membrane-dependent protein–protein interactions are important in membrane trafficking and cell signaling pathways. Cells employ many proteins, including small GTPases, cargo adaptors, and vesicle coats, that bind membranes to mediate the trafficking of cargo proteins by sculpting organelle membranes to generate transport vesicles and tubules. One prominent example is the small GTPase Arf1, a highly conserved regulator of membrane trafficking at the Golgi complex [1–3]. Upon activation at the membrane surface, Arf1 recruits many different effectors, including cargo adaptors and lipid modifying enzymes, to Golgi membranes where they carry out their function. Thus, Arf1 and its paralogs act as molecular switches to drive multiple Golgi membrane trafficking pathways.

William J. Brown (ed.), *The Golgi Complex: Methods and Protocols*, Methods in Molecular Biology, vol. 1496,
DOI 10.1007/978-1-4939-6463-5_4, © Springer Science+Business Media New York 2016

Some Arf1 effectors exhibit affinity for the membrane independent of binding to Arf1 [3–6]. One such effector is the exomer complex, which is responsible for sorting a subset of cargos at the *trans*-Golgi network (TGN) in the model organism *Saccharomyces cerevisiae* [4, 7–10]. To effectively assay for membrane binding and Arf1-dependent membrane recruitment, we employed a liposome pelleting assay where proteins and liposomes were incubated and subjected to centrifugation [6]. Protein bound to liposomes becomes enriched in the pellet fraction. The relative amount of the protein in the supernatant and pellet fractions was determined by quantification of band intensity after SDS-PAGE analysis. Thus, the relative efficiency of membrane binding and recruitment of different protein constructs and mutants can be measured quantitatively.

In addition to recruiting cargo adaptors necessary for trafficking, Arf1 and other related small GTPases, such as Sar1, have been implicated in membrane remodeling, defined as inducing membrane curvature or otherwise deforming the membrane [11–16]. Upon activation through GTP-binding, Arf1 inserts its myristoylated N-terminal amphipathic helix into the cytoplasmic leaflet of the membrane [17–19], and this membrane insertion is what drives membrane remodeling. Several other factors are known to play a role in membrane remodeling, including protein crowding, protein scaffolding, and protein shape [20, 21].

We were interested in determining what role, if any, Arf effectors such as the exomer complex might have in regulating or enhancing membrane curvature. Exomer is the only cargo adaptor known to sort proteins from the TGN to the apical plasma membrane, yet it only sorts ~1–5% of *S. cerevisiae* plasma membrane proteins and is not found in metazoans. Thus, we view Exomer as an important model system that should provide insight into how the bulk of proteins are sorted to the apical plasma membrane. Exomer is a heterotetramer consisting of two copies of Chs5 and any two members of four paralogous proteins known as the ChAPs (Chs5 and Arf1 binding proteins: Chs6, Bud7, Bch1, and Bch2). The ChAPs determine cargo specificity by binding directly to the cytoplasmic tails of cargo proteins, but appear to be structurally interchangeable within the complex [4, 6–8, 10, 22].

To investigate whether exomer can cooperate with Arf1 to remodel membranes, we employed a method similar to one previously reported by the McMahon lab [15], in which different sized liposomes are separated by differential sedimentation. Thus, the products of a vesiculation reaction (small liposomes) remain in the supernatant fraction, while the substrates (large liposomes) are found in the pellet fraction. We modified the established protocol by including a fluorescent tracer lipid in our liposome preparations, facilitating quantification of the liposomes.

Here we describe the methodology necessary for the preparation of liposomes and performing the Arf1/effector membrane binding and remodeling assays.

2 Materials

2.1 Preparation of Synthetic Liposomes

1. Rotary evaporator (*see* **Note 1**).
2. 25 ml glass pear-shaped flasks.
3. Liposome extruder and 400 nm pore filters (Avanti Polar Lipids).
4. Glass syringes, sizes from 10 to 500 μl.
5. Chloroform, ACS grade.
6. Methanol, ACS grade.
7. Lipid stocks (*see* recipes in Table 1).
    (a) Folch Fraction I.
    (b) Synthetic and natural lipids (Avanti Polar Lipids—*see* Table 1).
    (c) DiR lipid (Life Technologies) or other fluorescent lipid tracer (*see* **Note 2**).
8. HK buffer: 20 mM HEPES pH 7.4, 150 mM potassium acetate.

2.2 Liposome Pelleting/Membrane Binding Assay

1. Purified myristoylated Arf1 (for protocol, *see* ref. [23]) and purified Exomer protein constructs (*see* refs. [6, 24] for exomer purification method), or other protein to be used in the assay.
2. "TGN" liposomes (described in this protocol), or liposomes of another relevant composition.
3. HKM buffer: 20 mM HEPES pH 7.4, 150 mM potassium acetate, 1 mM magnesium chloride.
4. Beckman Coulter Optima TLX ultracentrifuge with a TLA 100.3 rotor.
5. 1.5 ml polyallomer tubes.
6. SDS sample (gel-loading) buffer.

2.3 Liposome Vesiculation Assay

1. Purified myristoylated Arf1 (described in Subheading 2.2, **item 1**), or other protein to be used in the assay.
2. Folch liposomes (described in this protocol).
3. HKM buffer: 20 mM HEPES pH 7.4, 150 mM potassium acetate, 1 mM magnesium chloride.
4. 0.5 M EDTA solution, pH 8.0.
5. 40 mM $MgCl_2$.
6. Guanine nucleotides (GDP and GTP, or the non-hydrolyzable GTP analogs GMP-PNP or GTPγS), 10 mM solutions in water.
7. Beckman Coulter Optima TLX ultracentrifuge with a TLA 100.3 rotor.
8. 1.5 ml polyallomer tubes.
9. SDS sample (gel-loading) buffer.

Table 1
Lipid composition formulations

"TGN" liposome lipid composition	
Mol%	*Lipid*
24	DOPC
6	POPC
7	DOPE
3	POPE
1	DOPS
2	POPS
1	DOPA
2	POPA
29	PI
1	PI(4)P
2	CDP-DAG
4	PO-DAG
2	DO-DAG
5	Ceramide (C18)
10	Cholesterol
1	DIR
Folch liposome lipid composition	
Mol%	*Lipid*
99	Folch Lipids (Folch Fraction I, Sigma)
1	DiR

Lipid formulations for the two types of liposomes employed in these methods. The "TGN" liposome formulation was selected based on a published lipidomics study of yeast TGN/early endosomal membranes [25]

3   Methods

3.1 Preparation of Synthetic Liposomes

Different liposome compositions may be employed in liposome pelleting and vesiculation assays, and the liposome composition should be customized depending upon the protein(s) being investigated. To assay membrane binding of exomer constructs, we generally use a lipid composition intended to mimic the yeast TGN (Table 1). In contrast, for the vesiculation assay, it appears necessary to use Folch lipids. The Folch liposomes can also be used for the binding assay.

1. Make 0.5–4 ml of an appropriate lipid master mix, with a final concentration of 2 mM lipids dissolved in 20:1 chloroform–methanol (*see* **Notes 3** and **4**). *See* Table 1 for master mix formulations.

2. Transfer 500 μl of the 2 mM lipid master mix to a 25 ml pear-shaped flask and dry on a rotary evaporator under vacuum, spinning the flask at an intermediate speed, for approximately 30 min at 30 °C or until all of the solvent has evaporated from the sample, resulting in a thin lipid film on the bottom of the flask (*see* **Notes 1** and **5**).

3. Rehydrate the lipids by gently adding 1 ml of HK buffer to the lipid film and incubating at 37 °C for 1–4 h.

4. Gently swirl or rotate the flask so that the lipid film slowly peels off the glass and large strands or ribbons of lipids can be seen, continue until the lipids have been completely resuspended.

5. Using a mini-extruder (Avanti Polar Lipids), extrude the lipids through a sizing filter, extruding back-and-forth 19 times. We use 400 nm pore size filters for both the membrane-binding (pelleting) and vesiculation experiments (*see* **Note 6**).

6. Transfer liposome solution to an Eppendorf or conical tube and store at 4 °C (*see* **Note 7**). The final concentration of lipids in the liposome solution is 1 mM.

3.2 Membrane Binding (Liposome Pelleting) Assay

This is a variation on the classical liposome pelleting assay, in which protein binding to membranes is monitored by measuring the fraction of the protein that pellets together with the liposomes. By measuring and subtracting the amount of protein that pellets in the absence of liposomes, we found that this assay can be quantified and yields results that are highly reproducible [6, 24]. We present two variations of this assay, one to measure Arf1-independent membrane binding, and the other to measure Arf1-dependent membrane binding. By utilizing both versions of this assay, one can dissect the relative contributions of intrinsic membrane binding versus Arf1-binding for membrane recruitment of an Arf1-effector. An example result of this experiment is shown in Fig. 1.

3.2.1 Arf1-Independent Membrane Binding (Liposome Pelleting) Assay

This protocol as detailed uses the Chs5/Chs6 exomer complex, but should be suitable for most membrane-binding proteins or protein complexes.

1. Mix liposomes (600 μM final lipid concentration) with Chs5(1-299)/Chs6 exomer complex (1 μM final concentration) in a final volume of 40 μl HKM buffer in a 1.5 ml polyallomer tube (*see* **Notes 8** and **9**). Gently mix by flicking the tube.

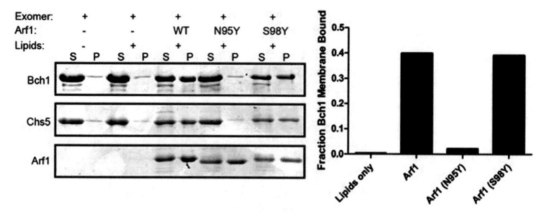

Fig. 1 Detecting membrane-dependent protein–protein interactions using a quantitative liposome pelleting assay. Liposome pelleting assay to determine Arf1-dependent membrane recruitment of the exomer complex, Chs5(1-299)/Bch1. Mutations in Arf1 (N95Y and S98Y) at the Arf1-exomer interface were made to assess the ability of Arf1 to interact with exomer and recruit it to the membrane. Presence of protein in the pellet (P) fraction indicates an interaction with Arf1-GTP. *S* supernatant, *P* pellet, *WT* wild-type. The *left panel* shows the gel, the *right panel* shows quantification. Band intensities were quantified using ImageJ. The fraction of Bch1 membrane bound was calculated using: $[P_{liposomes\ (+Arf1)}/(P_{liposomes\ (+Arf1)}+S_{liposomes\ (+Arf1)})]-[P_{no\ liposomes}/(P_{no\ liposomes}+S_{no\ liposomes})]$, where the amount of Bch1 pelleted in the absence of liposomes is subtracted from the amount of Bch1 pelleted in the presence of Arf1 and liposomes

2. Incubate at room temperature for 15 min. After 10 min gently mix by flicking the tube.

3. Spin the samples in Beckman TLX tabletop ultracentrifuge and TLA 100.3 rotor at 55,000 rpm (~150,000 × *g*) for 10 min at 4 °C (*see* **Note 10**).

4. Remove the supernatant to another tube and add 10 μl of 5× SDS sample buffer. Add 50 μl of 1× SDS sample buffer to the pellet.

5. Heat the samples to 55 °C for 15 min with frequent vortexing to mix, then run on SDS-PAGE. Do not run the dye-front off of the gel.

6. Stain the gels using a Coomassie dye solution lacking methanol or ethanol, or stain with LiCor IRDye colloidal Coomassie solution.

7. Scan the gels on an Odyssey imager (Li-COR). Alternatively, another scanner instrument can be used (*see* **Note 2**).

8. Determine the protein band intensities using ImageJ or the Odyssey software. Note the presence of lipids from the liposomes in the dye-front. It is not necessary to quantify lipids in this experiment, but it is necessary to verify that virtually all of the lipids are in the pellet fraction.

9. Calculate the percent protein pelleted according to this formula: $[P_{liposomes}/(P_{liposomes}+S_{liposomes})]-[P_{no\ liposomes}/(P_{no\ liposomes}+S_{no\ liposomes})]$, where P = pellet, S = supernatant, and "no liposomes" represents the background level of protein pelleted in the absence of liposomes (*see* **Note 11**).

A similar method can be used to determine the membrane-dependent protein–protein interaction between Arf1 and one of its effectors (such as exomer). For this protocol, Arf1 must first be activated (GTP-bound) in the presence of the liposomes to allow for membrane anchoring.

1. Mix liposomes (600 μM final lipid concentration) with Arf1 (7.5 μM final concentration), EDTA (1 mM final concentration), 125 μM GTP, GMP-PNP, or GTPγS (to activate Arf1) or GDP (as a negative control) in a final volume of 40 μl HKM buffer in a 1.5 ml polyallomer tube. Gently mix by flicking the tube (*see* **Note 12**).

2. Allow the exchange reaction to continue for 15–30 min at room temperature.

3. Stop the exchange reaction by adding 2 μl of $MgCl_2$ stock solution to a final concentration of 2 mM.

4. Add exomer or other Arf1 effector to the reaction at a final concentration of 1 μM and mix by flicking the tube.

5. Follow **steps 2–9** of the previous protocol (Subheading 3.2.1). Do not run the dye front off of the gel. Ensure that virtually all of the lipids are present in the pellet fractions and not in the supernatant fractions. It is possible to observe lipid vesiculation with certain proteins under certain conditions. The appearance of lipids in the supernatant fraction is suggestive of vesiculation (*see* Subheading 3.3).

**3.3 Liposome
Vesiculation Assay** It has been previously observed that small GTPases of the Arf family can directly induce membrane deformation through insertion of the N-terminal amphipathic helix into the membrane [11–16]. To understand what role exomer might play in Arf1 membrane remodeling activity, we modified a previously described experiment from the McMahon lab [15] where the ability to observe changes in liposome size via protein dependent membrane remodeling was monitored using high speed centrifugation to separate liposomes of different sizes: larger liposomes (400 nm diameter) pellet, while smaller liposomes (below 100 nm diameter) remain in the supernatant. The lipids are visualized near the dye-front of the SDS-PAGE gel, taking advantage of a fluorescent lipid incorporated into the liposomes.

Overall, the vesiculation assay protocol is similar to the Arf1-dependent membrane pelleting assay described above in Subheading 3.2.2, **step 2**, but there are important differences in the types of liposomes (*see* **Note 13**) and amounts of reagents used. We detail the protocol for the Chs5/Bch1 exomer complex, but other Arf1-dependent effectors can be investigated. The major strengths of this protocol are the ability to process many reactions in parallel and to quantitate the results. An example experimental result is shown in Fig. 2.

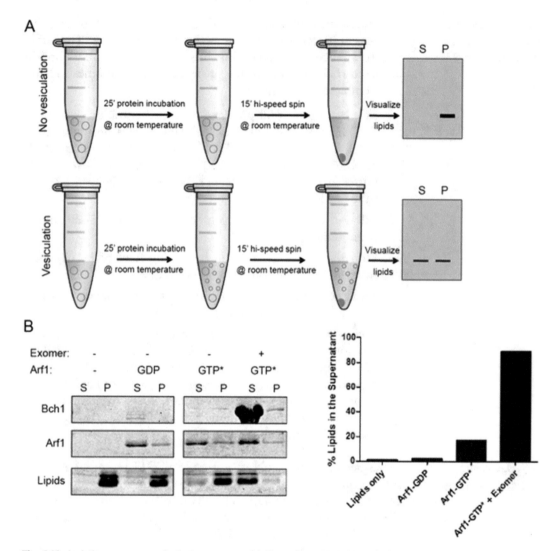

Fig. 2 Vesiculation assay reveals that exomer and Arf1 cooperate to remodel membranes. (**a**) Schematic of the membrane vesiculation assay. At high speeds, larger liposomes (~400 nm) will pellet completely, while smaller liposomes (~30 nm) will not pellet. A liposome population of mixed sizes can be spun at high speed and separated by size, with larger liposomes (and any bound proteins) in the pellet fraction and smaller liposomes remaining in the supernatant fraction. Lipids can be visualized via SDS-PAGE due to the presence of a fluorescent lipid dye in the liposomes. (**b**) Results of a membrane vesiculation assay to determine the role of exomer (Chs5(1-299)/Bch1) on Arf1-dependent membrane remodeling. The *left panel* shows the gel and the *right panel* shows quantification of lipids. Substrate liposomes pellet (P), while smaller liposomes generated from protein-dependent membrane remodeling remain in the supernatant (S). GTP* = GMP–PNP. Lipid band intensities were quantified using ImageJ and the percent of lipids found in the supernatant was calculated. Note that the high protein–lipid ratio results in a smaller fraction of total Arf1 binding to the membrane, and precise determination of the amount of Arf1 bound to membranes is confounded by vesiculation because Arf1 bound to smaller vesicles will remain in the supernatant

1. For "physiological" conditions, where the ratio of exomer to lipids and exomer to Arf1 is low (1:1250 and 1:25, respectively, *see* **Note 14**) combine Folch liposomes (250 μM final concentration, *see* **Note 14**) with Arf1 (5 μM final

concentration, *see* **Note 15**) and 125 μM GTP, GMP-PNP, or GTPγS (to activate Arf1) or GDP (as a negative control) in a final volume of 40 μl HKM buffer in a 1.5 polyallomer tube. Gently mix by flicking the tube (*see* **Note 16**).

2. Allow the exchange reaction to continue for 10 min at room temperature.

3. Stop the exchange reaction by adding 2 μl of MgCl$_2$ stock solution to a final concentration of 2 mM. Let the reaction incubate for an additional 5 min.

4. Add Chs5(1-299)/Bch1 exomer complex (200 nM final concentration) (*see* **Note 14**).

5. Let the reaction incubate 10 min at room temperature. Mix the reaction after 5 min and at the end of the incubation by flicking the tube.

6. Spin the samples at 20 °C for 15 min in Beckman TLX tabletop ultracentrifuge and TLA 100.3 rotor at 55,000 rpm ($\sim150,000 \times g$).

7. Follow **steps 4–7** from Subheading 3.2.1 above.

8. Measure the lipid band intensities using ImageJ or ODYSSEY software to determine the percentage of total lipids found in the supernatant: $S/(S+P) \times 100$. The presence of lipids in the supernatant indicates a reduction in the size of liposomes, and thus vesiculation, assuming that no lipids are observed in the supernatant fractions of the negative controls (*see* **Note 17**).

4   Notes

1. Alternatively, many researchers dry lipid mixes under a nitrogen stream.

2. DiR, or a similar fluorescent lipid tracer, should be added at 1% (mol basis) for visualization and quantification of liposomes. The fluorophore should be chosen based on the available instrumentation. DiR works well with the Odyssey (Li-Cor) scanner. Nonfluorescent lipids can also be detected in the visible spectrum after staining with Coomassie stain, but at lower sensitivity.

3. Avoid using plastic tubes, pipets, and pipet tips with organic solvent solutions for this and subsequent steps. Glass vials and small glass syringes (Hamilton) should be used to avoid contamination of the solutions with dissolved plastic material.

4. Lipid stocks in organic solvents should be stored tightly sealed at −20 °C or ideally at −80 °C.

5. To store excess master mix, remove the organic solvent by evaporation and store the dried lipid film in tightly sealed glass

vials at −80 °C. To use at a later date, dissolve in the same volume of 20:1 chloroform–methanol solution.

6. A 400 nm pore size is used to generate liposomes that are large enough to pellet and be used as substrates for vesiculation in the assays. It is important to note that extruding through pore sizes larger than 100 nm can result in liposomes that are multilamellar, rather than unilamellar. In the worst cases, this can lead to heterogeneity of lipid composition exposed on the outer surfaces of liposomes and inaccurate estimates of the amount of membrane surface available for protein binding. Drying the lipid mix in a pear-shaped flask on a rotary evaporator is intended to mitigate these effects by creating a very thin film prior to solubilization in aqueous buffer.

7. Some researchers use liposome preparations only on the same day they were prepared, to avoid complications arising from lipid oxidation or phase changes. However, we find that our liposome preparations can be stored for a few weeks at 4 °C without ill effects. It is important to determine whether the behavior of your liposome preparation changes after storage, by comparing results obtained with freshly prepared liposomes to results obtained using stored liposomes. Our practice is to prepare a large batch of liposomes at once (often 2–4 ml total) for use with large scale experiments performed over the course of 1 or 2 weeks. In our experience, this is preferable to preparing a fresh batch every day or even every few days, as we observe batch-to-batch variation in liposome behavior that is greater than variation in behavior observed after storage. We suspect this depends on the precise lipid mix and preparation method used, so it is best to test for your specific case.

8. The concentrations given should be considered a starting point. You may need to alter the concentrations of proteins and lipid to optimize for your protein of interest.

9. To account for any liposome independent (background) pelleting of your protein, a control in which no liposomes are included in the reaction should be used for each pelleting experiment.

10. We have observed differences in the pelleting behavior of certain liposome compositions. 400 nm TGN and Folch liposomes completely pellet under these conditions, whereas 400 nm liposomes made up entirely of DOPC may not pellet as effectively. In such cases, longer centrifugation times may help. Also, if liposomes are densely coated with bound proteins, they pellet more effectively. This fact should be considered when interpreting the results.

11. A key assumption to this analysis of the pelleting assay is that the same "background" amount of protein will pellet, presumably

due to the presence of protein aggregates, in the presence or absence of liposomes. In our experience, this assumption is valid. For any new investigation of a protein by this method, it is important to pay close attention to the amount of background pelleting. Some proteins may exhibit a high level of background pelleting (i.e., >40 % of the protein pellets in the absence of liposomes). This situation may be improved if the protein stock is centrifuged at high speed ($150,000 \times g$ for 20 min at 4 °C) to "pre-clear" aggregates (this is good general practice), prior to use in the membrane-binding reaction. For some proteins, this pre-clearing step does not prevent background pelleting and a liposome flotation experiment may be preferable. For a liposome flotation protocol, *see* ref. [23].

12. This step can be setup as several master mixes if different Arf1-effector constructs will be tested in parallel.

13. Vesiculation assays work best using liposomes prepared from Folch fraction I lipids. This lipid mix is currently the standard in the field for studying membrane remodeling, as Folch liposomes appear to be more amenable to remodeling.

14. The concentration of the various constituents in a vesiculation assay can have a profound effect on the level of vesiculation. To further enhance the effects of Arf1-dependent membrane remodeling, we adjusted the exomer–lipids and exomer–Arf1 ratios. Exomer was used at 2 μM final concentration and Folch liposomes were used at 50 μM final concentration, increasing the ratios of exomer–lipids from 1:1250 to 1:25 and exomer–Arf1 from 1:25 to 1:2.5.

15. This concentration of Arf1 results in a small but detectable amount of vesiculation in our hands, which is amplified by exomer. Higher concentrations of Arf1 result in higher levels of vesiculation.

16. Crucial negative controls for this assay include reactions with liposomes only, liposomes with Arf1-GDP, and liposomes with Arf1-GDP and exomer.

17. Ideally, vesiculation should be confirmed using alternative approaches. For example, imaging the vesiculation reactions by negative stain-EM or analyzing the size distribution of liposomes by dynamic-light scattering [15, 24].

Acknowledgements

We thank members of the Fromme lab for helpful discussions. This work was supported by NIH/NIGMS grant R01GM098621.

References

1. Gillingham AK, Munro S (2007) The small G proteins of the Arf family and their regulators. Annu Rev Cell Dev Biol 23:579–611
2. Donaldson JG, Jackson CL (2011) ARF family G proteins and their regulators: roles in membrane transport, development and disease. Nat Rev Mol Cell Biol 12:362–375
3. Cherfils J (2014) Arf GTPases and their effectors: assembling multivalent membrane-binding platforms. Curr Opin Struct Biol 29C:67–76
4. Wang CW, Hamamoto S, Orci L, Schekman R (2006) Exomer: a coat complex for transport of select membrane proteins from the trans-Golgi network to the plasma membrane in yeast. J Cell Biol 174:973–983
5. Heldwein EE, Macia E, Wang J, Yin HL, Kirchhausen T, Harrison SC (2004) Crystal structure of the clathrin adaptor protein 1 core. Proc Natl Acad Sci U S A 101:14108–14113
6. Paczkowski JE, Richardson BC, Strassner AM, Fromme JC (2012) The exomer cargo adaptor structure reveals a novel GTPase-binding domain. EMBO J 31:4191–4203
7. Sanchatjate S, Schekman R (2006) Chs5/6 complex: a multiprotein complex that interacts with and conveys chitin synthase III from the trans-Golgi network to the cell surface. Mol Biol Cell 17:4157–4166
8. Trautwein M, Schindler C, Gauss R, Dengjel J, Hartmann E, Spang A (2006) Arf1p, Chs5p and the ChAPs are required for export of specialized cargo from the Golgi. EMBO J 25:943–954
9. Ziman M, Chuang JS, Tsung M, Hamamoto S, Schekman R (1998) Chs6p-dependent anterograde transport of Chs3p from the chitosome to the plasma membrane in Saccharomyces cerevisiae. Mol Biol Cell 9:1565–1576
10. Barfield RM, Fromme JC, Schekman R (2009) The exomer coat complex transports Fus1p to the plasma membrane via a novel plasma membrane sorting signal in yeast. Mol Biol Cell 20:4985–4996
11. Aridor M, Fish KN, Bannykh S, Weissman J, Roberts TH, Lippincott-Schwartz J, Balch WE (2001) The Sar1 GTPase coordinates biosynthetic cargo selection with endoplasmic reticulum export site assembly. J Cell Biol 152:213–229
12. Lee MC, Orci L, Hamamoto S, Futai E, Ravazzola M, Schekman R (2005) Sar1p N-terminal helix initiates membrane curvature and completes the fission of a COPII vesicle. Cell 122:605–617
13. Beck R, Sun Z, Adolf F, Rutz C, Bassler J, Wild K, Sinning I, Hurt E, Brugger B, Bethune J, Wieland F (2008) Membrane curvature induced by Arf1-GTP is essential for vesicle formation. Proc Natl Acad Sci U S A 105:11731–11736
14. Krauss M, Jia JY, Roux A, Beck R, Wieland FT, De Camilli P, Haucke V (2008) Arf1-GTP-induced tubule formation suggests a function of Arf family proteins in curvature acquisition at sites of vesicle budding. J Biol Chem 283:27717–27723
15. Boucrot E, Pick A, Camdere G, Liska N, Evergren E, McMahon HT, Kozlov MM (2012) Membrane fission is promoted by insertion of amphipathic helices and is restricted by crescent BAR domains. Cell 149:124–136
16. Lundmark R, Doherty GJ, Vallis Y, Peter BJ, McMahon HT (2008) Arf family GTP loading is activated by, and generates, positive membrane curvature. Biochem J 414:189–194
17. Kahn RA, Randazzo P, Serafini T, Weiss O, Rulka C, Clark J, Amherdt M, Roller P, Orci L, Rothman JE (1992) The amino terminus of ADP-ribosylation factor (ARF) is a critical determinant of ARF activities and is a potent and specific inhibitor of protein transport. J Biol Chem 267:13039–13046
18. Goldberg J (1998) Structural basis for activation of ARF GTPase: mechanisms of guanine nucleotide exchange and GTP-myristoyl switching. Cell 95:237–248
19. Randazzo PA, Terui T, Sturch S, Fales HM, Ferrige AG, Kahn RA (1995) The myristoylated amino terminus of ADP-ribosylation factor 1 is a phospholipid- and GTP-sensitive switch. J Biol Chem 270:14809–14815
20. McMahon HT, Gallop JL (2005) Membrane curvature and mechanisms of dynamic cell membrane remodelling. Nature 438:590–596
21. Stachowiak JC, Brodsky FM, Miller EA (2013) A cost-benefit analysis of the physical mechanisms of membrane curvature. Nat Cell Biol 15:1019–1027
22. Rockenbauch U, Ritz AM, Sacristan C, Roncero C, Spang A (2012) The complex interactions of Chs5p, the ChAPs, and the cargo Chs3p. Mol Biol Cell 23:4402–4415
23. Richardson BC, Fromme JC (2015) Biochemical methods for studying kinetic regulation of Arf1 activation by Sec7. Methods Cell Biol 130:101–126
24. Paczkowski JE, Fromme JC (2014) Structural basis for membrane binding and remodeling by the exomer secretory vesicle cargo adaptor. Dev Cell 30:610–624

25. Klemm RW, Ejsing CS, Surma MA, Kaiser HJ, Gerl MJ, Sampaio JL, de Robillard Q, Ferguson C, Proszynski TJ, Shevchenko A, Simons K (2009) Segregation of sphingolipids and sterols during formation of secretory vesicles at the trans-Golgi network. J Cell Biol 185:601–612

Chapter 5

STEM Tomography Imaging of Hypertrophied Golgi Stacks in Mucilage-Secreting Cells

Byung-Ho Kang

Abstract

Because of the weak penetrating power of electrons, the signal-to-noise ratio of a transmission electron micrograph (TEM) worsens as section thickness increases. This problem is alleviated by the use of the scanning transmission electron microscopy (STEM). Tomography analyses using STEM of thick sections from yeast and mammalian cells are of higher quality than are bright-field (BF) images. In this study, we compared regular BF tomograms and STEM tomograms from 500-nm thick sections from hypertrophied Golgi stacks of alfalfa root cap cells. Due to their thickness and intense heavy metal staining, BF tomograms of the thick sections suffer from poor contrast and high noise levels. We were able to mitigate these drawbacks by using STEM tomography. When we performed STEM tomography of densely stained chloroplasts of *Arabidopsis* cotyledon, we observed similar improvements relative to BF tomograms. A longer time is required to collect a STEM tilt series than similar BF TEM images, and dynamic autofocusing required for STEM imaging often fails at high tilt angles. Despite these limitations, STEM tomography is a powerful method for analyzing structures of large or dense organelles of plant cells.

Key words Electron tomography, Scanning transmission electron microscopy, High-angle annular dark field detector, Signal-to-noise ratio, Golgi stacks

1 Introduction

Scanning transmission electron microscopy (STEM) is a variant of transmission electron microscopy (TEM) in which images are formed pixel-by-pixel with a finely focused electron beam that scans over the specimen. STEM has advantages over bright-field (BF) TEM when used for imaging biological samples stained with a heavy metal. The most common detector for STEM is the high-angle annular dark field (HAADF) detector that registers electrons scattered from the specimen. Because heavy metal atoms in the section steeply deflect electrons, these electrons are preferentially detected in the HAADF mode. STEM produces clearer micrographs from thicker sections than does BF TEM, and micrographs taken with an electron beam focused on a spot in the STEM mode have better signal-to-noise ratios than those taken

William J. Brown (ed.), *The Golgi Complex: Methods and Protocols*, Methods in Molecular Biology, vol. 1496,
DOI 10.1007/978-1-4939-6463-5_5, © Springer Science+Business Media New York 2016

with a broad electron beam covering the entire viewing area in the BF TEM mode.

Electron tomography (ET) reconstructs cellular volumes at nanometer-level resolution in three dimensions (3D) from tilt-series images that are captured from many different angles from 0 ° up to about 70 ° [1, 2]. Because of the thickness that electrons have to penetrate at high tilt angles, sections are usually less than 300 nm thick for ET with BF TEM [3]. Since STEM can capture higher quality images from thick sections than BF TEM, ET reconstructions benefit from collecting tilt series with STEM [4]. By adopting STEM, it has been possible to produce electron tomograms from 1-μm thick samples of yeast cells, malaria parasites, cultured mammalian cells, and primary cultures of isolated neuronal cells [2, 4–7]. In those applications, STEM tomograms exhibited better contrast and higher signal-to-noise ratios than tomograms reconstructed from BF TEM images. By adjusting angles of electron detection, Hohmann-Marriott et al. (2009) were able to acquire STEM tomograms with resolutions better than those from HAADF STEM data [7].

Using ET, high-resolution structures of Golgi stacks in mammalian, fungal, algal, and plant cells have been characterized [8–12]. Diameters of individual Golgi stacks range 500–1500 nm, which are larger than thicknesses of single sections (250–300 nm). This means that 3D reconstructions of an entire Golgi stack requires tomograms from consecutive sections. These serial tomograms are aligned and stacked into a coherent tomogram that encompasses the cisternal stack, associated vesicles, surrounding Golgi matrix, and functionally linked organelles such as the endoplasmic reticulum. For example, a dividing Golgi stack in an *Arabidopsis* meristem cell was 1.8-mm wide, spanning seven serial sections [3, 13].

Golgi stacks in plant cells produce cell wall polysaccharides and proteoglycans. In mucilage-secreting root cap cells, polysaccharide synthesis in the Golgi stacks is activated. Golgi stacks in those cells are as large as 2500 nm and have hypertrophied peripheral vesicles [14]. It is not easy to reconstruct a comprehensive 3D structure of such an enlarged Golgi stack by stacking tomograms of serial sections. Furthermore, information about the 30- to 40-nm thick volumes between adjacent tomograms is lost during sectioning. Given that STEM tomography can produce tomograms of subcellular organelles from thicker sections, it is an attractive alternative for tomographic reconstruction of Golgi stacks in root cap cells [5].

We have performed HAADF STEM tomography to analyze Golgi stacks in 500-nm thick sections of alfalfa root cap cells. In agreement with the previous studies, signal-to-noise and membrane contrast were improved in STEM tomograms when compared with BF TEM tomograms. Data collection took approximately four times longer when the specimen was scanned at pixel sizes that match BF TEM resolutions. Dynamic focusing frequently failed at high angles, making it even slower to capture a tilt series. However, membranous structures with weak contrast such as endoplasmic reticulum and thylakoids were resolved more clearly in STEM tomograms than in BF TEM tomograms. This enhanced signal-to-noise ratio of cellular membranes facilitates automatic segmentation of membranous structures.

2 Materials

2.1 High Pressure Freezing

1. High-pressure freezer Leica HPM100 (Leica Microsystems, Buffalo Grove, IL, USA).

2. Specimen carrier: Aluminum type A and type B planchettes (Technotrade International, Manchester, NH, USA).

3. Liquid nitrogen.

4. Alfalfa (*Medicago sativa*) seeds.

5. Cryoprotectant: 0.15 M sucrose solution in distilled water.

2.2 Freeze Substitution

1. 2 % OsO_4 in acetone: Dissolve 1 g of OsO_4 in 50 ml anhydrous acetone. Aliquot into cryovials and freeze in liquid nitrogen. Frozen cryovials can be stored in liquid nitrogen.

2. AFS2 automatic freeze substitution machine (Leica Microsystems, Buffalo Grove, IL, USA).

2.3 Embedding

1. Embed-812 kit (Electron Microscopy Sciences, Hatfield, PA, USA). To prepare 40 g of resin, mix 20 g of Embed 812, 14 g of nadic methyl anhydride, and 6 g of dodecyl succinic anhydride; 1 g of benzyldimethylamine is added to the 40 g of resin mix to catalyze polymerization.

2. Flat embedding silicone mold (Cat. Nu. 70901, Electron Microscopy Sciences, Hatfield, PA, USA).

2.4 Ultramicrotomy and Post-staining

1. Ultramicrotome.

2. Formvar-coated copper slot grids (1×2 mm slot size).

3. Diamond knife.

4. 2 % (w/v) uranyl acetate (UA) in 70 % methanol and Reynolds' lead citrate solution.

2.5 Acquisition of Tilt Series and Tomographic Reconstruction

1. Intermediate voltage transmission electron microscope equipped with a motorized eucentric goniometer for collecting tilting series and STEM capability (*see* **Note 1**).

2. Grid holder for dual-axis tomography; for example, the Fischione Model 2040 holder (Fischione, Export, PA, USA).

3. A personal computer running a Linux or Unix operating system or a Windows computer with a Linux emulator such as Cygwin.

3   Methods

3.1   High-pressure Freezing and Freeze-Substitution

1. Scatter alfalfa (*Medicago sativa*) seeds on a piece of paper towel wetted with distilled water in a plastic petri dish. Place the petri dish in a plant growth chamber for seedlings to germinate. It helps to grow seedlings at a 60 ° angle so that roots grow straight.

2. After growing 3–4 days, dissect seedling root tips with a scalpel dipped in distilled water with 0.15 M sucrose. Put dissected tips in the B type planchette and cover with another B type planchette (*see* **Note 2**).

3. Freeze with the HPM100 freezer and transfer frozen samples into the cryovials with OsO_4 freeze-substitution medium.

4. Incubate cryovials with samples at –80 °C for 24 h and slowly raise the temperature to –20 °C (in 1 °C/h steps).

5. After 12 h at –20 °C, move the cryovials to 4 °C and incubate for 4 h. Leave the samples at room temperature for 1 h.

3.2   Embedding and Sectioning

1. After washing with anhydrous acetone three times, begin embedding in the Embed 812 resin mix.

2. After incubation in the 10 % dilution, raise the concentration of the Embed 812 resin from 25 to 100 % in increments of 25 %. At each dilution step, incubate the root specimens for several hours. Change 100 % resin three times, and transfer samples to a silicone mold.

3. Cure resin in a 60 °C oven for 2 days.

4. Mounting, sectioning, and post-staining should be carried out as described previously Kang (2010) [15].

5. Post-staining and preparation of slot grids for collecting tilt series should be performed as explained in Toyooka and Kang (2013) [16] (Fig. 1a, b).

Fig. 1 Transmission electron microscopy and electron tomography images of Golgi stacks in mucilage-secreting alfalfa root cap cells. (**a**) Low magnification electron micrograph of a longitudinal section through the root cap region of an alfalfa seedling. Root cap cells and the meristem zone are marked. Scale bar indicates 20 μm. (**b**) A cluster of Golgi stacks (G) in an alfalfa root cap cell. This image was acquired from a 500-nm thick section. V indicates vacuole. Scale bar indicates 500 nm. (**c, d**) A HAADF-STEM image of a 500 nm thick section from an alfalfa root cap cell (**c**) and its inverted image using the densnorm command (**d**). (**e, f**) Tomographic slice images of a cluster of Golgi stacks. E was taken from a BF tomogram of a 500-nm thick section. F was captured from a STEM tomogram of the same 500-nm thick sections imaged shown in (**c**). The BF tomogram in (**e**) was prepared from the same section examined by STEM in (**c**). Scale bars in (**c**–**f**) indicate 500 nm. (**g, h**) Higher magnification views of mucilage-carrying vesicles (*arrowheads*) in (**g**) a BF tomographic slice and (**h**) a STEM tomographic slice. (**i**) Outlines of the two vesicles shown in the panel (**h**) were traced by the auto-contour command of the IMOD software. Pixel sizes of the BF tomogram and the STEM tomogram are 1.50 and 1.81 nm, respectively (*see* **Note 7**). Scale bars in (**g**–**i**) indicate 50 nm. Scale bars indicate 50 nm

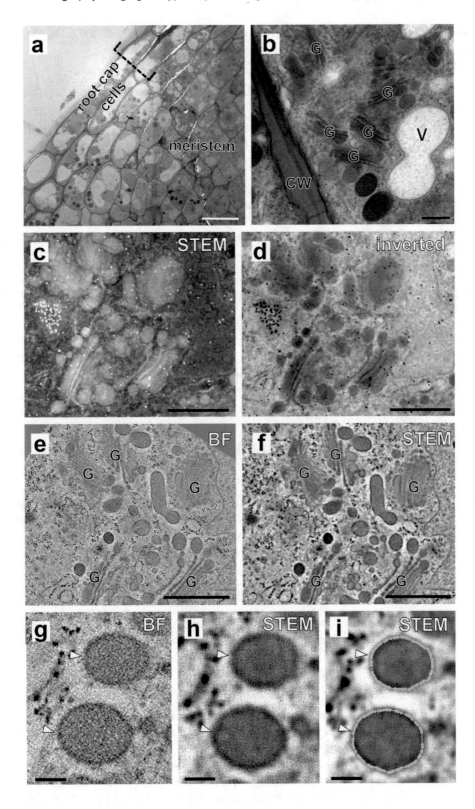

3.3 Acquisition and Tomogram Reconstruction

1. Load the grid into a TEM equipped with a STEM module and a HAADF detector.

2. Load STEM alignment and do direct alignment.

3. In the BF TEM mode, find the area of interest. Adjust eucentric height.

4. Go back to the STEM model, and open the STEM tomography program.

5. Collect tilt series (±60 ° with 1.5 ° intervals) (*see* **Notes 3–5**).

6. Dark-field images of the tilt series should be converted into bright-field images with the densenorm command (*see* **Note 6**, Fig. 1c, d). Tomograms should be calculated from the inverse contrasted tilt series as described by Toyooka and Kang (2013) [16].

4 Notes

1. We used an F20 intermediate voltage TEM from FEI (Hillsboro, OR, USA) and the STEM data acquisition software developed by the manufacturer.

2. We remove the border cells by gently shaking root tips in distilled water. The organelle integrity is compromised in these border cells.

3. Before starting data acquisition, check image brightness and contrast at 60 °. Brightness in high tilt images can be significantly different from brightness of the 0 ° image.

4. The exposure time we use for capturing tilt images is 20 μs per pixel. For tracking and eucentric centering, images are taken with the setting of 1.5 μs per pixel. It takes about 3 h to acquire 81 STEM images at 2048 by 2048 pixels (±60 ° with 1.5 ° interval including tracking, eucentric centering, and focusing at each angle). For a dual axis tomogram, two tilt series are required from a cellular volume.

5. The autofocusing function often fails at high tilt angles and data acquisition is interrupted. If that happens, focus manually and resume the acquisition process.

6. Options that should be selected for the densnorm command are "densnorm –su –log 0 –reverse inputfile outputfile".

7. The vesicle membranes are better resolved in STEM than in BF tomograms (Fig. 1e, f). Secretory vesicles and ribosomes appear darker in STEM, too. The bottom row of Fig. 1 shows vesicles in the BF tomogram (g), STEM tomogram (h), and STEM tomogram after auto tracing the vesicle surfaces (i). The auto tracing failed for the vesicles in the BF tomogram.

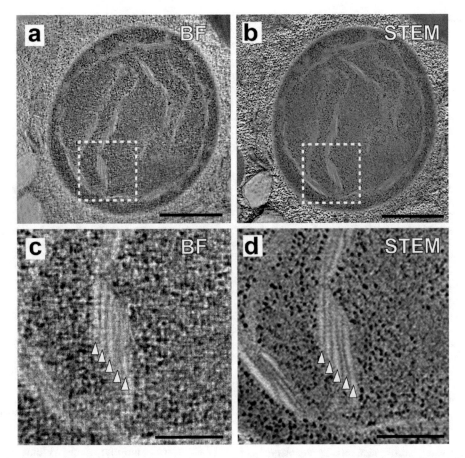

Fig. 2 A chloroplast in an *Arabidopsis* cotyledon examined by (**a** and **c**) BF tomography and (**b** and **d**) STEM tomography. Boxed areas in (**a**) and (**b**) are shown at higher magnification in (**c**) and (**d**), respectively. Scale bars in (**a**) and (**b**) indicate 1 μm and those in (**c**) and (**d**) indicate 250 nm. Raw tilt series were collected from a 400-nm thick section with a 200-KV transmission electron microscope. The five layers (marked with *arrowheads*) of a granum stack are clearly identified in d but the distinction is not so obvious in (**c**) (*see* **Note 8**). Pixel sizes of the BF tomogram and the STEM tomogram are 1.3 and 2.6 nm, respectively. The chloroplast samples were prepared as described in Lee et al. (2012) [17]

8. The chloroplast stroma is darker than the cytosol. Because of the strong staining, thylakoid membranes are not clearly discernable especially in images captured at high angles (>50 °). This drawback leads to weak contrast in BF tomograms (Fig. 2).

Acknowledgements

This work was supported by the Cooperative Research Program for Agriculture Science and Technology Development (Project No. PJ010953092015) from the Rural Development Administration, Republic of Korea and the Area of Excellence grant (AoE/M-05/12) from the Research Grants Council of Hong Kong.

References

1. Staehelin LA, Kang B-H (2008) Nanoscale architecture of endoplasmic reticulum export sites and of Golgi membranes as determined by electron tomography. Plant Physiol 147:1454–1468

2. Yakushevska AE, Lebbink MN, Geerts WJC et al (2007) STEM tomography in cell biology. J Struct Biol 159:381–391

3. Donohoe B, Mogelsvang S, Staehelin L (2006) Electron tomography of ER, Golgi and related membrane systems. Methods 39:154–162

4. Aoyama K, Takagi T, Hirase A, Miyazawa A (2008) STEM tomography for thick biological specimens. Ultramicroscopy 109:70–80

5. Sousa AA, Azari AA, Zhang G, Leapman RD (2011) Dual-axis electron tomography of biological specimens: extending the limits of specimen thickness with bright-field STEM imaging. J Struct Biol 174:107–114

6. Murata K, Esaki M, Ogura T et al (2014) Whole-cell imaging of the budding yeast Saccharomyces cerevisiae by high-voltage scanning transmission electron tomography. Ultramicroscopy 146:39–45

7. Hohmann-Marriott MF, Sousa AA, Azari AA et al (2009) Nanoscale 3D cellular imaging by axial scanning transmission electron tomography. Nat Methods 6:729–731

8. Donohoe BS, Kang B-H, Gerl MJ et al (2013) Cis-Golgi cisternal assembly and biosynthetic activation occur sequentially in plants and algae. Traffic 14:551–567

9. Donohoe BS, Kang B-H, Staehelin LA (2007) Identification and characterization of COPIa- and COPIb-type vesicle classes associated with plant and algal Golgi. Proc Natl Acad Sci U S A 104:163–168

10. Ladinsky MS, Mastronarde DN, McIntosh JR et al (1999) Golgi structure in three dimensions: functional insights from the normal rat kidney cell. J Cell Biol 144:1135–1149

11. Mogelsvang S, Gomez-Ospina N, Soderholm J et al (2003) Tomographic evidence for continuous turnover of Golgi cisternae in Pichia pastoris. Mol Biol Cell 14:2277–2291

12. Kang B-H, Nielsen E, Preuss ML et al (2011) Electron tomography of RabA4b- and PI-4Kβ1-labeled trans Golgi network compartments in Arabidopsis. Traffic 12:313–329

13. Kang B-H, Staehelin LA (2008) ER-to-Golgi transport by COPII vesicles in Arabidopsis involves a ribosome-excluding scaffold that is transferred with the vesicles to the Golgi matrix. Protoplasma 234:51–64

14. Staehelin L, Giddings T, Kiss J (1990) Macromolecular differentiation of Golgi stacks in root tips of Arabidopsis and Nicotiana seedlings as visualized in high pressure frozen and freeze-substituted samples. Protoplasma 157: 75–91

15. Kang B-H (2010) Electron microscopy and high-pressure freezing of Arabidopsis. Methods Cell Biol 96:259–283

16. Toyooka K, Kang B-H (2013) Reconstructing plant cells in 3D by serial section electron tomography. In: Žárský V, Cvrčková F (eds) Plant cell morphogenesis. Humana Press, Totowa, pp 159–170

17. Lee K-H, Park J, Williams DS et al (2013) Defective chloroplast development inhibits maintenance of normal levels of abscisic acid in a mutant of the Arabidopsis RH3 DEAD-box protein during early post-germination growth. Plant J 73:720–732

Chapter 6

Reconstitution of COPI Vesicle and Tubule Formation

Seung-Yeol Park, Jia-Shu Yang, and Victor W. Hsu

Abstract

The Golgi complex plays a central role in the intracellular sorting of proteins. Transport through the Golgi in the anterograde direction has been explained by cisternal maturation, while transport in the retrograde direction is attributed to vesicles formed by the coat protein I (COPI) complex. A more detailed understanding of how COPI acts in Golgi transport is being achieved in recent years, due in large part to a COPI reconstitution system. Through this approach, the mechanistic complexities of COPI vesicle formation are being elucidated. This approach has also uncovered a new mode of anterograde transport through the Golgi, which involves COPI tubules connecting the Golgi cisternae. We describe in this chapter the reconstitution of COPI vesicle and tubule formation from Golgi membrane.

Key words COPI, ARF, Golgi complex, Vesicle formation, Tubule formation

1 Introduction

Membrane traffic in the cell occurs in two general directions, exocytic (outward bound) and endocytic (inward bound). In the exocytic direction, proteins that are initially synthesized at the endoplasmic reticulum (ER) transit through the Golgi (in the anterograde direction) for sorting to other parts of the cell. In the endocytic direction, proteins at the plasma membrane also pass through the Golgi (in the retrograde direction) before reaching the ER. Thus, because the Golgi acts at the crossroad of the two general directions of transport, considerable interest exists in understanding how bidirectional transport through the Golgi complex is accomplished [1, 2].

For many years, anterograde Golgi transport has been explained by the movement of the Golgi cisternae, known as cisternal maturation, while retrograde Golgi transport has been thought to be mediated by vesicles formed by the COPI complex [1, 2]. Early studies identified coatomer, a multimeric complex, as the core components of the COPI complex [3, 4]. The small GTPase ARF1 was then identified to regulate coatomer by dictating its distribution

William J. Brown (ed.), *The Golgi Complex: Methods and Protocols*, Methods in Molecular Biology, vol. 1496,
DOI 10.1007/978-1-4939-6463-5_6, © Springer Science+Business Media New York 2016

between the functional (on membrane) and nonfunctional (cytosolic) pools [5]. Moreover, key upstream regulators of ARF1 were also identified and/or characterized. These include a guanine nucleotide exchange factor (GEF) activity that catalyzes the activation of ARF1 [6, 7], and a GTPase-activating protein (GAP) that catalyzes ARF1 deactivation [8].

Subsequently, our group has been making major contributions to the further understanding of COPI transport [9]. We initially elucidated a more complex role for the GAP that catalyzes ARF1 deactivation in COPI transport, known as ARFGAP1. Whereas early studies predicted that ARFGAP1 acts in the uncoating of COPI vesicles [10], we found that it has a novel function in promoting COPI vesicle formation, with mechanistic elucidation suggesting that ARFGAP1 acts as another component of the COPI complex [11, 12]. We then identified BARS (Brefeldin-A ADP-Ribosylated Substrate) to act at the fission stage of COPI vesicle formation [13, 14]. Further elucidating how vesicle fission occurs, we identified a key lipid, phosphatidic acid (PA), which promotes the ability of BARS to bend membranes in achieving vesicle fission [15]. Notably, rather than acting merely to recruit BARS to membrane, we found that PA participates actively with BARS to bend membranes [15].

We then uncovered an even more complex role for the COPI complex by finding that it also generates tubules that connect the Golgi stacks [16]. Mechanistically, this involves COPI coupling with distinct lipid enzymatic activities in dictating whether Golgi membranes form vesicles or tubules [16]. More recently, we have defined that transport through COPI tubules complements cisternal maturation in explaining how anterograde Golgi transport occurs [17]. A notable mechanistic detail is that the small GTPase Cdc42 plays a pivotal role in coordinating bidirectional Golgi transport by targeting the two major functions of COPI—carrier formation and cargo sorting [17].

These achievements have been due, in large part, to a reconstitution system that allows novel factors to be identified, as well as elucidating how they act. The reconstitution system is divided into two stages (summarized in Fig. 1). In the first stage, purified ARF1 and coatomer are incubated with Golgi membrane to achieve the recruitment of coatomer onto membrane. In the second stage, ARFGAP1 and BARS are added, which then result in COPI vesicle formation. To achieve COPI tubule formation, either cytoplasmic phospholipase type 2 alpha (cPLA$_2\alpha$) or Cdc42 is added additionally in the second stage, which results in vesicle formation being diverted to tubule formation. We describe below the technical details of these reconstitutions.

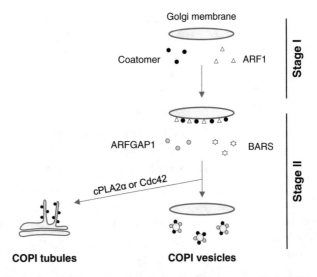

Fig. 1 Schematic of the two-stage incubation system to reconstitute COPI vesicles and tubules. In the first stage, Golgi membrane is incubated with purified ARF1 and coatomer. In the second stage, the incubated Golgi membrane is further incubated with purified ARFGAP1 and BARS to form COPI vesicles. For COPI tubule formation, the second-stage incubation involves the further addition of either purified cPLA$_2$α or Cdc42

2   Materials

2.1 Recombinant Proteins Produced from Bacteria

1. 50 μM sodium myristate: dissolve 12.52 mg/l culture media.

2. 1 M isopropyl β-D-1-thiogalactopyranoside (IPTG): dissolve 238 g/ml distilled water.

3. 1 M DL-dithiothreitol (DTT): dissolve 154 mg/ml distilled water.

4. Protease inhibitor cocktail.

5. ARF1 lysis buffer: 20 mM Tris pH 8.0, 100 mM NaCl, 1 mM MgCl$_2$, 1 mM DTT, 1 mg/ml lysozyme, and protease inhibitor.

6. Columns: HiTrap Q HP, HiPrep 26/60 Sephacryl S-100, HiTrap phenylsepharose (GE Healthcare Life Sciences) and Ni-NTA resin (Clontech).

7. HiTrap Q HP running buffer: 20 mM Tris pH 8.0, 100 mM NaCl, 1 mM MgCl$_2$, 1 mM DTT, and 10% glycerol.

8. HiPrep 26/60 Sephacryl S-100 running buffer: 20 mM Tris pH 8.0, 100 mM NaCl, 1 mM MgCl$_2$, and 1 mM DTT.

9. High-salt buffer: 20 mM Tris pH 8.0, 5 M NaCl, 1 mM MgCl$_2$, and 1 mM DTT.

10. HiTrap phenylsepharose running buffer: 20 mM Tris pH 8.0, 3 M NaCl, 1 mM $MgCl_2$, and 1 mM DTT.

11. Reaction buffer: 25 mM HEPES pH 7.2, 50 mM KCl, 2.5 mM $Mg(OAc)_2$, and 1 mM DTT.

12. BARS lysis buffer: 20 mM Tris pH 8.0, 100 mM NaCl, 1 mg/ml lysozyme, 1% Triton X-100, and protease inhibitor.

13. BARS washing buffer: 50 mM Tris pH 7.5, 100 mM NaCl, 5 mM $MgCl_2$, and 20 mM imidazole.

14. BARS elution buffer: 50 mM Tris pH 7.5, 100 mM NaCl, 5 mM $MgCl_2$, and 250 mM imidazole.

2.2 Recombinant Proteins Produced from Insect Cells

1. BestBac 2.0 Baculovirus cotransfection kit (Expression systems).

2. ESF 921 insect cell culture medium, protein-free (Expression systems).

3. Bac-to-Bac Baculovirus Expression system (Invitrogen).

4. ARFGAP1 lysis buffer: 20 mM Tris pH 7.5, 100 mM NaCl, 1% Triton X-100, 1% CHAPS, and protease inhibitor.

5. ARFGAP1 washing buffer: 50 mM Tris pH 7.5, 200 mM NaCl, 5 mM $MgCl_2$, 0.1% CHAPS, and 20 mM imidazole.

6. ARFGAP1 elution buffer: 50 mM Tris pH 7.5, 200 mM NaCl, 5 mM $MgCl_2$, 0.1% CHAPS, and 250 mM imidazole.

7. Hypotonic buffer: 20 mM sodium borate pH 10.2, 5 mM $MgCl_2$, and protease inhibitor.

8. TBS-M: 50 mM Tris pH 7.2, 150 mM NaCl, 5 mM $MgCl_2$, and protease inhibitor.

9. Cdc42 washing buffer: 50 mM Tris pH 7.2, 500 mM NaCl, 5 mM $MgCl_2$, 0.1% CHAPS, and 20 mM imidazole.

10. Cdc42 elution buffer: 50 mM Tris pH 7.2, 500 mM NaCl, 5 mM $MgCl_2$, 0.1% CHAPS, and 250 mM imidazole.

2.3 Purification of Coatomer from Rat Liver

1. Rat liver (freshly prepared in the lab).

2. Polytron homogenizer (7 mm generator diameter, Kinematica AG).

3. 2 mM ethylenediaminetetraacetic acid (EDTA): dissolve 0.58 mg/ml distilled water.

4. Homogenize buffer: 25 mM Tris pH 8.0, 500 mM KCl, 250 mM sucrose, 2 mM EDTA, 1 mM DTT, and protease inhibitor.

5. Saturated ammonium sulfate.

6. Dialysis buffer: 25 mM Tris pH 7.5, 200 mM KCl, 1 mM DTT, and protease inhibitor.

7. 0.45 μm syringe filter.

8. Columns: DEAE-Sepharose FF, HiTrap Q HP and Resource Q (GE Healthcare Life Sciences).

9. Running buffer: 25 mM Tris pH 7.5, 200 mM KCl, 1 mM DTT, and 10 % glycerol.

10. Anti-βCOP antibody (M3A5).

11. Dilution buffer: 25 mM Tris pH 7.5, 1 mM DTT, and 10 % glycerol.

2.4 Golgi Membrane Preparation

1. Cell culture media: RPMI1640 supplemented with 10 % FBS, 2 mM L-glutamine, 20 mM HEPES, and 20 μg/ml gentamicin.

2. ST buffer: 856 mg of sucrose/ml 10 mM Tris–HCl, pH 7.4.

3. Ball-bearing homogenizer and 25 μm clearance ball [18].

4. Sucrose gradient buffer: 62 % (805 mg sucrose/ml 10 mM Tris, pH 7.4), 35 % (403 mg sucrose/ml with 10 mM Tris, pH 7.4), and 29 % (325 mg sucrose/ml 10 mM Tris, pH 7.4).

5. Refractometer.

6. 30 ml syringe, 10 ml syringe.

7. 50 ml ultracentrifuge tubes for SW-28 rotor (Beckman).

2.5 In Vitro Reconstitution

1. Low binding microcentrifuge tubes and tips.

2. Traffic buffer: 25 mM HEPES pH 7.2, 50 mM KCl, 2.5 mM Mg(OAc)$_2$, 1 mg/ml soybean trypsin inhibitor, 1 mg/ml BSA, and 200 mM sucrose.

3. 3 M KCl solution: dissolve 224 mg KCl/ml traffic buffer.

4. BSA solution: dissolve 0.1 g BSA in 10 ml of traffic buffer.

5. Cushion buffer: 25 mM HEPES pH 7.2, 50 mM KCl, 2.5 mM Mg(OAc)$_2$, and 15 % sucrose.

2.6 Electron Microscopy

1. Formvar carbon coated grids.

2. 2 % paraformaldehyde (PFA) in PBS (from 16 % stock solution).

3. Uranyl acetate staining solution: carefully mix 1 ml of 4 % uranyl acetate and 9 ml of 2 % methyl cellulose.

4. JEOL 1200EX transmission electron microscope.

3 Methods

3.1 Generating Recombinant Proteins Using Bacterial Expression

3.1.1 Purification of Myristoylated-ARF1

1. Culture BL21 cells containing Arf1/pET3 and *N*-myristoyltransferase/pBB131 plasmids in 1 l LB media supplemented with 100 μg/ml ampicillin and 25 μg/ml kanamycin at 37 °C.

2. Add 50 μM sodium myristate when O.D. is 0.6, and further culture the cells for 30 min.

3. Induce the expression using 0.1 mM of IPTG at room temperature, overnight.

4. Harvest the cells using centrifugation at $6000 \times g$, 4 °C for 10 min.

5. Lyse the cells using 50 ml of lysis buffer for 30 min, followed by sonication.

6. Centrifuge cell lysate at $100,000 \times g$, 4 °C for 1 h.

7. Load the supernatant to HiTrap Q HP column, which is equilibrated with running buffer at 1 ml/min.

8. Pool the fractions containing ARF1 determined by 15% SDS-PAGE.

9. Concentrate purified ARF1 using Amicon Ultra-15 (MWCO: 10,000) at $2000 \times g$ until 10 ml of volume.

10. Load the sample to HiPrep 26/60 Sephacryl S-100 column, which is equilibrated with running buffer, and further develop the column using running buffer at 2 ml/min.

11. Pool the fractions containing ARF1 determined by 15% SDS-PAGE.

12. Adjust salt concentration to 3 M using high-salt buffer (*see* **Note 1**).

13. Load the sample to HiTrap phenylsepharose, which is equilibrated with running buffer at 0.5 ml/min.

14. Wash the column using 20 ml of running buffer.

15. Elute myristoylated-ARF1 by decreasing salt concentration to 100 mM in 15 ml.

16. Pool fractions containing myristoylated-ARF1, as determined by 15% SDS-PAGE.

17. Dialyze against reaction buffer at 4 °C, overnight.

18. Measure protein concentration using Bradford assay and stored at −80 °C.

3.1.2 Purification of BARS

1. Culture BL21 cells containing BARS/pET-15b plasmid in 1 l LB media supplemented with 100 μg/ml ampicillin at 37 °C.

2. When O.D. is 0.6, induce the expression using 0.1 mM IPTG for 3 h.

3. Harvest the cells using centrifugation at $6000 \times g$, 4 °C for 10 min.

4. Lyse the cells using 20 ml of lysis buffer at 4 °C for 30 min, followed by sonication.

5. Centrifuge cell lysate at $30,000 \times g$ at 4 °C for 30 min.

6. Load the supernatant to 2 ml of Ni-NTA column.

7. Wash the column with 100 ml of washing buffer.

8. Elute recombinant BARS using elution buffer.

9. Pool fractions containing BARS, as determined by 10% SDS-PAGE.

10. Dialyze against reaction buffer at 4 °C, overnight.

11. Measure protein concentration using Bradford assay and stored at −80 °C.

3.2 Generating Recombinant Proteins Using Baculovirus Expression

3.2.1 Purification of ARFGAP1 and cPLA2α

1. Use ArfGAP1 and cPLA2α in pVL1392 plasmids to prepare baculovirus using BestBac 2.0 Baculovirus co-transfection kit.

2. Infect 1:50 diluted P2 virus to 4.2×10^6 Sf9 cells/ml (2 l) at 27 °C.

3. Harvest the cells before viability drops below 80%.

4. Centrifuge at $500 \times g$ for 20 min.

5. Lyse the cells using 100 ml of lysis buffer at 4 °C for 1 h.

6. Centrifuge cell lysate at $6000 \times g$ for 20 min.

7. Load the supernatant to 2 ml of Ni-NTA column.

8. Wash the column using washing buffer.

9. Elute recombinant proteins using elution buffer.

10. Pool fractions containing recombinant proteins determined by 15% SDS-PAGE.

11. Dialyze against reaction buffer at 4 °C, overnight.

12. Measure protein concentration using Bradford assay, and stored at −80 °C.

3.2.2 Purification of Prenylated Cdc42

1. Use cdc42 in pFASTBAC THb plasmid to prepare Baculovirus infected insect cells (BIICS) using Bac-to-Bac Baculovirus Expression system.

2. Infect 1 ml of BIICS to 1.5×10^6 Sf9 cells/ml (2 l) at 27 °C.

3. Harvest the cells before viability drops below 80%.

4. Centrifuge at $500 \times g$ for 20 min.

5. Lyse the cells using 100 ml of hypotonic buffer, followed by using Dounce homogenizer (20 times).

6. Centrifuge cell lysate at $150,000 \times g$ for 30 min.

7. Resuspend the pellet using 100 ml of TBS-M.

8. Centrifuge at $150,000 \times g$ for 30 min.

9. Solubilize prenylated Cdc42 in the pellets using 50 ml of TBS-M containing 1% Triton X-100 at 4 °C for 1 h.

10. Centrifuge at $9000 \times g$ for 20 min.

11. Load the supernatant to 2 ml of Ni-NTA column.

12. Wash the column using washing buffer.

13. Elute prenylated Cdc42 using elution buffer.

14. Pool fractions containing prenylated Cdc42 determined by 15 % SDS-PAGE.

15. Dialyze against reaction buffer at 4 °C, overnight.

16. Measure protein concentration using Bradford assay, and stored at −80 °C until use.

3.3 Purification of Coatomer from Tissue

1. Homogenize 100 g of fresh rat liver in 30 ml of homogenize buffer using polytron homogenizer (*see* **Note 2**).

2. Centrifuge at 10,000×*g* for 30 min.

3. Take the supernatant for second centrifugation at 41,000×*g* for 1 h.

4. Dilute the supernatant with homogenize buffer to give 7 mg/ml.

5. Add saturated ammonium sulfate to the homogenate to give 35 % ammonium sulfate concentration. Mix at 4 °C for 20 min.

6. Centrifuge at 7500×*g* for 10 min.

7. Resuspend the pellet using 40 ml of dialysis buffer.

8. Dialyze against dialysis buffer at 4 °C, overnight.

9. Centrifuge at 10,000×*g* for 10 min.

10. Take the supernatant for second centrifugation at 41,000×*g* for 45 min.

11. Filtrate the supernatant through 0.45 μm syringe filter.

12. Load the sample to DEAE-Sepharose FF, which is equilibrated with running buffer at 1 ml/min.

13. Elute the proteins up to 1 M KCl by a linear gradient in 600 ml.

14. Pool the fractions containing coatomer determined by western blotting using anti-βCOP antibody.

15. Dilute the fractions with dilution buffer to give 200 mM KCl.

16. Load the mixture to HiTrap Q HP column, which is equilibrated with running buffer.

17. Elute coatomer using running buffer containing 500 mM KCl.

18. Pool the fractions containing the peak of protein.

19. Mix the sample with dilution buffer to give 200 mM KCl.

20. Load the sample to Resource Q column at 1 ml/min.

21. Elute coatomer using KCl gradient up to 1 M in 60 ml.

22. Pool the fractions containing coatomer, determined by western blotting using anti-βCOP antibody.

23. Dialyze against reaction buffer at 4 °C, overnight.

24. Measure protein concentration using Bradford assay, and stored at −80 °C until use.

3.4 Golgi Membrane Preparation

1. Culture CHO cells in four roller bottles (250 ml media/bottle) until confluent.

2. Collect the cells using trypsin treatment, followed by centrifugation at $500 \times g$ for 10 min.

3. Resuspend the pellet using cold ST buffer (10 ml ST buffer/ml pellet).

4. Centrifuge at $500 \times g$ for 10 min.

5. Resuspend the pellet using four times volume of ST buffer.

6. Homogenize the cells using ball-bearing homogenizer with 22 μm ball for 10–12 passes (*see* **Note 3**).

7. Centrifuge the sample at $2000 \times g$ for 10 min.

8. Mix 12 ml of supernatant with 11 ml of 62 % sucrose buffer and 230 μl of 100 mM EDTA, pH 7.4 (*see* **Note 4**).

9. Prepare sucrose gradient. Place 9 ml of 29 % sucrose buffer and 15 ml of 35 % sucrose buffer from the bottom of ultracentrifuge tubes using 30 ml syringe with blunt tip needle.

10. Gently load 12 ml of sample mixture at the bottom of sucrose gradient using 30 ml syringe with blunt tip needle.

11. Centrifuge at $110,000 \times g$ for 2.5 h.

12. Collect Golgi membrane at the 29 %/35 % sucrose interface using 10 ml syringe with 18G needle gauge.

13. Measure protein concentration and then store in a liquid N_2 tank.

3.5 COPI Reconstitution System

1. Resuspend Golgi membrane (250 μg/200 μl) in low binding microcentrifuge tube using 1 ml of Traffic buffer, followed by centrifugation at $15,000 \times g$ for 30 min.

2. Take out supernatant.

3. Wash Golgi membrane using 1 ml of 3 M KCl on ice–water bath for 5 min.

4. Centrifuge at $15,000 \times g$ for 30 min.

5. Take out supernatant.

Stage I

6. Resuspend the pellets with 100 μl of traffic buffer and then add 50 μl of BSA solution.

7. Incubate the mixture at 37 °C for 20 min.

8. Add ARF1 (6 μg/ml), coatomer (6 μg/ml), and GTP (2 mM).

9. Make up to 500 μl with traffic buffer.

10. Incubate the mixture on 37 °C water bath for 15 min.

11. Stop reaction on ice–water bath for 5 min.

12. Place 10 µl of cushion buffer at the bottom of the tube (*see* **Note 5**).

13. Centrifuge at $15,000 \times g$ for 20 min.

14. Take out supernatant.

Stage II

15. Gently resuspend the pellet using traffic buffer (*see* **Note 6**).

16. Add ARFGAP1 (6 µg/ml) and BARS (3 µg/ml) to generate COPI vesicles, or additionally with Cdc42 (6 µg/ml) and/or cPLA2α (3 µg/ml) to generate COPI tubules.

17. Make up to 50 µl with traffic buffer, followed by incubation at 37 °C water bath for 20 min.

18. Stop reaction on ice–water bath for 5 min.

19. Place 5 µl of cushion buffer at the bottom of the tube.

20. Centrifuge at $15,000 \times g$ for 10 min.

21. Gently collect the supernatant and pellet (*see* **Notes 7** and **8**).

22. Determine COPI carrier formation by western blotting using anti-βCOP antibody.

3.6 Electron Microscopy (EM)

EM analysis is performed to determine reconstituted COPI transport carriers (vesicle and tubule). To examine cargo sorting into the carriers, immunogold labeling approach is performed. In this case, Golgi membrane extracted from the cells expressing VSVG-Myc or VSVG-KDELR-Myc was used.

1. Load the sample (**step 18** in Subheading 3.5) on grids for 10 min.

2. Fix the sample using 2 % PFA/PBS for 10 min.

3. Incubate the grids using 1 % BSA/PBS for 10 min.

4. Rinse the grids 7× using distilled water for 2 min.

5. Incubate the sample with uranyl acetate staining solution for 10 min.

6. Examine the reconstituted COPI transport carriers using TEM at 80 kV.

Immunogold labeling

7. Incubate the grids (**step 3** in Subheading 3.6) with mouse anti-Myc antibody (9E10) for 1 h.

8. Rinse the grids 3× using PBS for 5 min.

9. Incubate the grids using a rabbit anti-mouse antibody for 30 min.

10. Rinse the grids 3× using PBS for 5 min.

11. Incubate the grids using protein A conjugated with 10 nm gold particle for 30 min.

12. Rinse the grids 2× using PBS for 5 min.

13. Rinse the grids 7× using distilled water for 2 min.

14. Incubate the sample with uranyl acetate staining solution for 10 min.

15. Examine cargoes in the reconstituted COPI transport carriers using TEM at 80 kV.

4   Notes

1. Mix the sample with 5 M NaCl solution by the ratio between 2.94 of 5 M NaCl and 2.06 of 0.1 M NaCl.

2. Homogenize 10 g of tissue in each time.

3. Check how much cells were homogenized using Trypan Blue staining. 10% cells are usually homogenized by 12 passes.

4. Check mixture of sucrose concentration, which should be about 37%, using a refractometer.

5. 15% sucrose in cushion buffer separates Golgi membrane from other components.

6. Vigorous pipetting may cause artificially generated vesicles.

7. Touching pellet might cause inconsistent results.

8. Supernatant contains a mixture of soluble proteins and vesicles, pellet contains Golgi membranes.

Acknowledgements

We thank Jian Li for discussions. This work was funded by grants from the National Institutes of Health to V.W.H. (R37GM058615), and also by the Basic Science Research Program of the National Research Foundation of Korea to S.-Y.P. (2014R1A6A3A030 56673).

References

1. Glick BS, Nakano A (2009) Membrane traffic within the Golgi apparatus. Annu Rev Cell Dev Biol 25:113–132

2. Nakano A, Luini A (2010) Passage through the Golgi. Curr Opin Cell Biol 22:471–478

3. Malhotra V, Serafini T, Orci L, Shepherd JC, Rothman JE (1989) Purification of a novel class of coated vesicles mediating biosynthetic protein transport through the Golgi stack. Cell 58:329–336

4. Waters MG, Serafini T, Rothman JE (1991) 'Coatomer': a cytosolic protein complex containing subunits of non-clathrin-coated Golgi transport vesicles. Nature 349:248–251

5. Donaldson JG, Cassel D, Kahn RA, Klausner RD (1992) ADP-ribosylation factor, a small GTP-binding protein, is required for binding of the coatomer protein beta-COP to Golgi membranes. Proc Natl Acad Sci U S A 89: 6408–6412

6. Donaldson JG, Finazzi D, Klausner RD (1992) Brefeldin A inhibits Golgi membrane-catalysed exchange of guanine nucleotide onto ARF protein. Nature 360:350–352

7. Helms JB, Rothman JE (1992) Inhibition by brefeldin A of a Golgi membrane enzyme that catalyses exchange of guanine nucleotide bound to ARF. Nature 360:352–354

8. Cukierman E, Huber I, Rotman M, Cassel D (1995) The ARF1 GTPase-activating protein: zinc finger motif and Golgi complex localization. Science 270:1999–2002

9. Hsu VW, Lee SY, Yang J-S (2009) The evolving understanding of COPI vesicle formation. Nat Rev Mol Cell Biol 10:360–364

10. Tanigawa G, Orci L, Amherdt M, Ravazzola M, Helms JB, Rothman JE (1993) Hydrolysis of bound GTP by ARF protein triggers uncoating of Golgi-derived COP-coated vesicles. J Cell Biol 123:1365–1371

11. Lee SY, Yang J-S, Hong W, Premont RT, Hsu VW (2005) ARFGAP1 plays a central role in coupling COPI cargo sorting with vesicle formation. J Cell Biol 168:281–290

12. Yang JS, Lee SY, Gao M, Bourgoin S, Randazzo PA, Premont RT, Hsu VW (2002) ARFGAP1 promotes the formation of COPI vesicles, suggesting function as a component of the coat. J Cell Biol 159:69–78

13. Yang J-S, Lee SY, Spano S, Gad H, Zhang L, Nie Z, Bonazzi M, Corda D, Luini A, Hsu VW (2005) A role for BARS at the fission step of COPI vesicle formation from Golgi membrane. EMBO J 24:4133–4143

14. Yang J-S, Zhang L, Lee SY, Gad H, Luini A, Hsu VW (2006) Key components of the fission machinery are interchangeable. Nat Cell Biol 8:1376–1382

15. Yang J-S, Gad H, Lee SY, Mironov A, Zhang L, Beznoussenko GV, Valente C, Turacchio G, Bonsra AN, Du G, Baldanzi G, Graziani A, Bourgoin S, Frohman MA, Luini A, Hsu VW (2008) A role for phosphatidic acid in COPI vesicle fission yields insights into Golgi maintenance. Nat Cell Biol 10:1146–1153

16. Yang JS, Valente C, Polishchuk RS, Turacchio G, Layre E, Moody DB, Leslie CC, Gelb MH, Brown WJ, Corda D, Luini A, Hsu VW (2011) COPI acts in both vesicular and tubular transport. Nat Cell Biol 13:996–1003

17. Park S-Y, Yang J-S, Schmider AB, Soberman RJ, Hsu VW (2015) Coordinated regulation of bidirectional COPI transport at the Golgi by CDC42. Nature 521:529–532

18. Balch WE, Dunphy WG, Braell WA, Rothman JE (1984) Reconstitution of the transport of protein between successive compartments of the Golgi measured by the coupled incorporation of N-acetylglucosamine. Cell 39:405–416

Chapter 7

Reconstitution of Phospholipase A$_2$-Dependent Golgi Membrane Tubules

Edward B. Cluett, Paul de Figueiredo, Marie E. Bechler,
Kevin D. Thorsen, and William J. Brown

Abstract

The Golgi complex is the Grand Central Station of intracellular membrane trafficking in the secretory and endocytic pathways. Anterograde and retrograde export of cargo from the Golgi complex involves a complex interplay between the formation of coated vesicles and membrane tubules, although much less is known about tubule-mediated trafficking. Recent advances using in vitro assays have identified several cytoplasmic phospholipase A$_2$ (PLA$_2$) enzymes that are required for the biogenesis of membrane tubules and their roles in the functional organization of the Golgi complex. In this chapter we describe methods for the cell-free reconstitution of PLA$_2$-dependent Golgi membrane tubule formation. These methods should facilitate the identification of other proteins that regulate this process.

Key words Golgi complex, Membrane tubules, In vitro reconstitution, PLA$_2$ enzymes

1 Introduction

Export of cargo from the Golgi complex and *trans* Golgi network (TGN) is mediated by membrane vesicles and membrane tubules. COPI and clathrin AP-1-coated vesicles are well characterized for their roles in retrograde and anterograde trafficking, respectively, from the Golgi and TGN [1, 2]. Other "uncoated" vesicles, including AP-3, AP-4, and exomer vesicles, also mediate sorting and export from the TGN [3]. In addition to membrane vesicles, membrane tubules, ~60–80 nm in diameter and varying lengths, have been shown to export cargo from the Golgi complex, including retrograde transport from the *cis* Golgi to the endoplasmic reticulum (ER) and anterograde trafficking from the TGN to the plasma membrane [4, 5]. In addition, membrane tubules have been shown to mediate intra-cisternal movement of cargo molecules [6] and to link the mammalian Golgi complex into an intact ribbon-like organelle [7, 8]. However, the exact functions of

William J. Brown (ed.), *The Golgi Complex: Methods and Protocols*, Methods in Molecular Biology, vol. 1496,
DOI 10.1007/978-1-4939-6463-5_7, © Springer Science+Business Media New York 2016

membrane tubules, and their mechanisms of formation are less understood than coated vesicles. Currently, two mechanisms of Golgi membrane tubule formation have been described: COPI-dependent [9] and COPI-independent [4, 5] tubulation. The focus of this chapter is on COPI-independent Golgi membrane tubules.

COPI-independent membrane tubules were revealed when Lippincott-Schwartz and colleagues discovered the remarkable properties of the fungal metabolite brefeldin A (BFA), which induces exaggerated membrane tubule formation from all Golgi compartments and TGN membranes [10–13], as well as endosomes [11, 13, 14]. BFA inhibits the activation of the small GTPase Arf1, which is used to recruit COPI and AP-1 clathrin subunits on Golgi membranes [15, 16]. As a consequence of the loss of Arf1 and its associated coats, Golgi and TGN membranes form extensive membrane tubule networks. This work led to a reevaluation of Golgi morphology, which resulted in the realization that studies from as early as the 1960s reported extensive membrane tubules emanating from the Golgi complex [17, 18]. These observations were confirmed by more modern tomographic reconstructions from electron micrographs [19–21] and by live cell imaging of fluorescently tagged organelle markers [22, 23]. Thus, membrane tubules are normal features of Golgi complexes and the TGN, for example as in Fig. 1, but they remained unappreciated for many years.

Reasoning that BFA-induced membrane tubules are formed by an unknown, COPI-independent mechanism, we sought to identify the molecules involved in this process. One of the first steps was to demonstrate COPI-independent formation of Golgi membrane tubules in a cell-free, reconstitution system [24–26]. Using this in vitro assay, and the in vivo BFA model system, we discovered through the use of small molecule inhibitors that Golgi and TGN membrane tubules require the action of a cytoplasmic Ca^{2+}-independent phospholipase A_2 (PLA$_2$) enzyme [7, 8, 27]. Through biochemical and molecular genetic approaches, we discovered that the cytoplasmic Ca^{2+}-independent PLA$_2$ enzyme Platelet Activating Factor Acetylhydrolase (PAFAH) Ib is involved in the biogenesis of Golgi and TGN membrane tubules [28]. Other labs later discovered additional PLA$_2$ enzymes that contribute to Golgi and ER-Golgi-Intermediate Compartment (ERGIC) membrane tubule formation, respectively, cPLA$_2$α [29] and PLA2G6 [30]. Thus, the Golgi complex as a whole requires the enzymatic activity of a surprising set of PLA$_2$ enzymes to produce membrane tubules. Remarkably, a similar requirement for PLA$_2$ enzyme activity for the production of endosome membrane tubules has been revealed [31–34].

These studies have generated much interest and many new questions about membrane tubules in general [4], and Golgi membrane tubules, in particular. First, the extent to which these

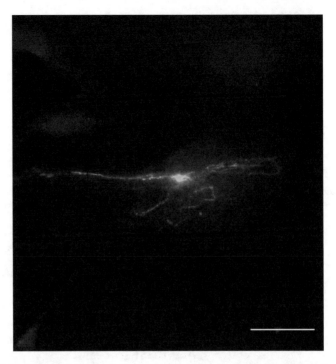

Fig. 1 Demonstration of membrane tubules extending from the TGN. Cells were expressing gp135-EGFP from the RUSH system, which was developed by Boncompain et al. [42]. The RUSH system allows for the inducible and synchronous release of any GFP-tagged secretory or membrane protein from the ER, which can be followed by fluorescence microscopy. In this case, the trafficking of gp135-EGFP (*green signal*), which is a plasma membrane protein, was imaged 45 min after release from the ER, a time at which it is found in membrane tubules emanating from the TGN. The TGN was counterstained by immunofluorescence using anti-TGN46 antibodies and rhodamine labeled secondary antibody (*red signal*)

enzymes play unique or redundant roles is unclear. Second, there is little information about how these enzymes are regulated in time and space to ensure the appropriate levels of membrane tubule formation in response to secretory load. Third, the molecular mechanisms by which the PLA$_2$ enzymes change the shape of Golgi and TGN membranes into thin tubular structures is unknown, although several provocative models have been proposed [4, 5, 35].

The answers to these questions will be greatly aided by the in vitro, reconstitution assay that we have developed and which is described below. Three main procedures are described: (1) preparation of cytosolic extracts; (2) isolation of intact Golgi complexes; and (3) in vitro tubulation of Golgi membranes and visualization by negatively stained EM samples. Previous studies on Golgi membrane tubule formation revealed that bovine brain cytoplasm (BBC) is a convenient source of "tubulation factors", i.e., cytoplasmic PLA$_2$ enzymes [24, 26]. Thus, BBC serves as a useful positive control and likely source of other proteins that regulate the process of membrane tubule formation. However, other sources of cytoplasmic extracts can also be used [26]. Also, we have used rat liver as a convenient source of intact Golgi complexes, but cultured cells can be used as well [9, 36].

2 Materials

2.1 Preparation of Bovine Brain Cytosol

1. Phenylmethanesulfonylfluoride (PMSF) stock: 10 mg/ml in isopropanol; store in aliquots at –20 °C. Warm the aliquots to room temperature immediately before use.

2. Storage buffer: 320 mM sucrose, 25 mM Tris–HCl, pH 7.4.
   - recommended volume: 2 l. Store at 4 °C.

3. Homogenization buffer: 25 mM Tris–HCl, 500 mM KCl, 250 mM sucrose, 1 mM dithiothreitol (DTT), with the following (final concentrations) added from stock solutions: 2 mM EGTA, 2 μg/ml aprotinin, 0.5 μg/ml leupeptin, 2 μM pepstatin A, 0.5 mM 1,10-phenanthroline. Adjust to pH 7.4. Add PMSF to 1 mM.
   - recommended volume: 1.5 l. Store at 4 °C.

4. Saturated ammonium sulfate: approximately 716 g into 1 l, pH 8.0 by use of NaOH pellets.

5. Dialysis buffer: 25 mM Tris–HCl, 50 mM KCl, pH 8.0.
   - Prepare with and without 0.5 mM DTT (*see* below).
   - recommended volumes: 1 l with DTT, 4 l without DTT.

6. Sorvall centrifuge with GSA rotor or equivalent.

7. Floor model ultracentrifuge with Beckman Type 35 or 50.2ti fixed angle rotor.

8. Waring blender or equivalent.

9. Prechilled centrifuge tubes, several graduated cylinders, and beakers.

2.2 Isolation of Intact Golgi Complexes from Rat Liver

1. 100 mM Tris–HCl, pH 7.4.

2. 2.3 M sucrose (store at 4 °C).

3. ST buffer (0.25 M sucrose/Tris): 7.56 ml 2.3 M sucrose, 7 ml 100 mM Tris, QS to 70 ml with ddH$_2$O.

   *QS (latin for quantum sufficiat, i.e., a sufficient quantity; therefore, to QS a sample to 70 ml means to bring that sample to 70 ml final volume).

4. Sucrose step gradient solutions.

   Prepare on the day of use or 1 day before use, and QS all with 4 °C ddH$_2$O. The recipe provides sufficient volume for six

Beckman SW28 centrifuge tubes or equivalent. Store solutions at 4 °C or on ice. The sucrose steps are as follows: 0.8 M sucrose/Tris (10.4 ml of 2.3 M sucrose, 3 ml of 100 mM Tris, QS to 30 ml); 0.9 M sucrose/Tris (11.7 ml of 2.3 M sucrose, 3 ml of 100 mM Tris, QS to 30 ml); 1.0 M sucrose/Tris (21.5 ml of 2.3 M sucrose, 5 ml of 100 mM Tris, QS to 50 ml); 1.2 M sucrose/Tris (26 ml of 2.3 M sucrose, 5 ml of 100 mM Tris, QS to 50 ml).

5. Surgical instruments for small animal dissection including fine scissors and forceps.

6. Balch–Rothman ball bearing homogenizer [37] with a clearance of 0.0054″. For our homogenizer, we used a 0.2460″ ball (*see* **Note 1**).

7. 10 ml disposable syringes that fit securely into the homogenizer.

8. Floor model ultracentrifuge with Beckman SW28 swinging bucket rotor (or equivalent).

9. Single-edge razor blades.

10. 150–200 g CD Sprague-Dawley rat (*see* **Note 2**).

11. 150 ml beaker.

12. Beckman Sorvall centrifuge with SS-34 rotor and 15 ml Corex tubes (or equivalent).

13. Phase contrast microscope with slides and cover glass.

2.3 In Vitro Golgi Membrane Tubule Formation Reaction and EM Negative Stain

1. 10 mM ATP (store aliquots at −20 °C; only thaw once).

2. 100 mM MgCl$_2$ (store at −20 °C).

3. 10 mM HEPES, pH 7.4 (store at 4 °C).

4. Reaction buffer: 25 mM Tris, pH 8.0, 50 mM KCl; store at room temperature (RT).

5. Rat liver Golgi suspension (store aliquots at −80 °C; thaw only once, keep on ice, and use immediately).

6. 2 % phosphotungstic acid, pH 7.4 (pH with NaOH and stored at 4 °C).

7. Formvar- and carbon-coated EM grids.

8. Self-closing fine forceps.

9. Kimwipes.

10. Transmission EM.

2.4 Immunogold Labeling of Negatively Stained Golgi Membranes

1. Periodate-lysine-paraformaldehyde fixative [38]: 2 % paraformaldehyde, 0.75 M lysine, 10 mM NaIO$_4$ in 35 mM phosphate buffer, pH 6.2.

2. EM-grade primary antibodies against Golgi-localized protein of interest.

3. Gold-labeled secondary antibodies (usually 5 or 10 nm gold-labeled antibodies).

4. Formvar- and carbon-coated nickel EM grids.

5. Buffer A: 0.1 % ovalbumin in PBS, pH 7.4.

6. Buffer B: Buffer A containing 0.05 % Triton X-100.

3   Methods

3.1   Preparation of Bovine Brain Cytosol (BBC)

This procedure uses bovine brains as a source of Golgi membrane tubulation factors. This procedure is modified from others designed to induce the formation of COPI-coated vesicles from Golgi membranes [39]. However, as modified here, this BBC preparation will only support membrane tubule formation, not COPI vesicle formation [24, 26]. Obtain intact bovine brains from slaughterhouse or other supplier (*see* **Note 3**). The method described below uses 300 g of cerebral cortex as starting material, and will yield ~80 ml of BBC. All procedures are carried out in a cold room at 4 °C and/or on ice.

1. Immerse brain material in storage buffer for transport from slaughterhouse (~300 ml for 300 g brain).

2. Remove blood and meninges, discarding dura mater, cerebellum, and tissue other than the cerebrum.

3. Weigh out 300 g of prepared brain and place in pan or tray for processing.

4. Measure out 750 ml of homogenization buffer (add 1.5 ml of PMSF stock just before use). Pour 50 ml over the brain and into the tray.

5. Using a new single-edge razor blade, mince the brain into ~2 mm pieces.

6. Using the remaining homogenization buffer, transfer the minced brain into a Waring blender.

7. Homogenize with two 30 s bursts.

8. Centrifuge the homogenate in 250 ml capped bottles in a Sorvall GSA rotor at $9200 \times g$ for 45 min to pellet nuclei and extracellular debris.

9. Carefully pour and collect the post-nuclear supernatant (PNS) (yield ~650 ml).

10. Measure the PNS volume and add PMSF from stock (2 μl PMSF/ml of PNS).

11. Prepare a "crude cytosol" fraction by centrifugation in a Beckman Type 35 rotor at $142,400 \times g$ max for 2.25 h (or equivalent centrifugation with other rotors, e.g., Beckman 50.2ti rotor).

12. Carefully harvest the clarified crude cytosol with 10 ml pipette and automatic pipettor, being careful to not disrupt the pellets. Yield is ~400 ml.

13. Precipitate proteins with 60 % ammonium sulfate as follows:
    – over the course of 1 h, add saturated ammonium sulfate with a peristaltic pump, mixing with a magnetic stir bar (600 ml of saturated ammonium sulfate/400 ml of crude cytosol). Stir 1 h more. (*See* **Note 4**).

14. Pellet the precipitated proteins by centrifugation in a Sorvall GSA rotor at ~12,000×*g* max for 45 min.

15. Discard the supernatant.

16. Dissolve the pellets in each bottle in 10–20 ml of dialysis buffer containing DTT. Use a 10 ml pipet and automatic pipettor. Dissolving the pellets will require vigorous pipetting, but avoid making air bubbles (which are problematic for later dialysis).

17. Collect all of the dissolved material and measure volume in a cold graduated cylinder.

18. Add 2 μl of PMSF stock/ml of dissolved protein solution; carefully mix.

19. Dialyze in 12–14 kDa cutoff dialysis tubing against 4 l of dialysis buffer (+DTT) for 1 h (gently mixing with magnetic stir bar).

20. Dialyze as in **step 19** but with dialysis buffer (–DTT) overnight (gently stirring buffer with a magnetic stir bar, making sure to avoid abrasion to the dialysis tubing).
    – a precipitate will form during this time.

21. Harvest the dialysate into a cold graduated cylinder and add 2 μl of PMSF stock/ml of dialysate; gently mix.

22. Centrifuge the dialysate in a Beckman 50.2ti rotor at 245,000×*g* max for 1 h.

23. Collect the supernatant and add 2 μl of PMSF stock/ml of supernatant. This preparation is the "bovine brain cytosol" (BBC) that will be used in membrane tubulation assays.

24. Measure the protein concentrations with Bradford or other protein assay. Final concentration of protein should be ~25 mg/ml.

25. Store in 0.5–1 ml aliquots at –80 °C in cryotubes. BBC will remain active for ~6 months when stored at –80 °C; however, do not refreeze a thawed aliquot.

3.2 Isolation of Intact Golgi Complexes from Rat Liver

This method uses a Balch–Rothman ball bearing homogenizer [37], which was originally designed for cultured mammalian cells, to break open liver cells. We have found that it is far superior to Potter-Elvehjem, Dounce-type, or other homogenizers for preparing intact Golgi complexes from rat or other animal organs.

1. Sacrifice rat with CO_2 using approved institutional protocol for animal euthanasia (*see* **Note 2**).

   All subsequent steps are done on ice or at 4 °C in cold room.

2. Excise liver and weigh. Best results are obtained with livers <10 g, but all sizes can be used. Removing the connective tissue is more time consuming with larger livers.

3. Cut away any connective tissue, blood vessels, etc. These will interfere with homogenization. It is important to remove as much connective tissue as possible. The success of the prep depends on smooth passage through the homogenizer. Small livers have less connective tissue, and therefore smoother passage. Rinse two to three times with ST buffer to remove blood.

4. Mince liver into a fine puree with razor blade. Add 1 ml ST buffer during the mincing. It helps to flip the liver over to ensure pieces are cut through entire depth. The finer the mincing, the better the homogenization will go; however, excessive mincing can lead to damaged cells.

5. Make a 20 % weight to volume suspension of pureed liver in ST buffer, e.g., 10 g liver + 50 ml ST buffer. Put the resultant material in a cold 150 ml beaker on ice.

6. Add ST buffer with a transfer pipette or syringe to the homogenization chamber to remove all air bubbles.

7. Swirl to liver pieces to uniformly suspend and quickly load into a 10 ml syringe (homogenization will be done in 10 ml batches). Liver puree should be easily drawn into the syringe. If not, draw it into the syringe and push it out several times. Invert the syringe and remove all air bubbles before homogenization.

8. Homogenize with the Balch/Rothman homogenizer (*see* **Note 1**). Three passes for each batch are usually sufficient. More than three passes usually causes intact Golgi complexes to fragment and cisternae to detach. If the homogenizer jams, stop immediately, remove the ball bearing from the homogenizer, and clear away the connective tissue. Reassemble the homogenizer, add ST buffer to displace air, and proceed. Pool homogenates.

9. Inspect homogenate by phase contrast microscopy to ensure ~70 % cells breakage. Look for intact nuclei that are not surrounded by plasma membranes. Over-homogenization can lead to ruptured nuclei and DNA release, which can affect the quality of the Golgi isolation.

10. Centrifuge homogenate at 3000×g for 10 min at 4 °C in Sorvall SS-34 rotor (use 15 ml Corex tubes) to generate post-nuclear supernatants.

11. Collect and pool the post-nuclear supernatants (PNS) and keep on ice. Check by phase contrast microscopy to ensure that nuclei are depleted from the PNS. *See* comment for **step 9** above.

12. Add an equal volume of 2.3 M sucrose to the pooled PNS and gently mix.

13. Carefully layer sucrose step gradients in Beckman SW28.1 ultracentrifuge tubes, starting with PNS at the bottom of the tube (using solutions described in Subheading 2.2, **step 4**). Pipette very slowly to avoid disruption of the interfaces/mixing of the layers (*see* **Note 5**): 15 ml of PNS/2.3 M sucrose/Tris, 8 ml of 1.2 M sucrose/Tris, 8 ml of 1.0 M sucrose/Tris, 4 ml of 0.9 M sucrose/Tris, 3–4 ml of 0.8 M sucrose/Tris.

14. Balance tubes with 0.8 M sucrose/Tris by dropwise addition to the top of each gradient.

15. Centrifuge at 120,000×*g* max for 2.5 h with slow acceleration and brake off.

16. Collect intact Golgi complexes at the 0.9/1.0 M sucrose interface. There will be milky white bands at each interface (Fig. 2). First, carefully remove the gradient above the 0.9/1.0 interface by vacuum-assisted aspiration with a Pasteur pipet. Gently move the pipet in a circular motion from the periphery to the center and back, repeating this until all of the solution above the 0.9/1.0 M interface has been removed. Gently harvest the milky 0.9/1.0 Golgi-enriched layer with a Pasteur pipet. As much as possible, avoid removing the layer below.

17. Aliquot into 1.5 ml microfuge tubes (200 μl/tube) and freeze at –80 °C.

3.3 Cell-Free, In Vitro Reconstitution of Golgi Membrane Tubules

This assay promotes the formation of Golgi membrane tubules, but not COPI-coated vesicles. It was developed to rapidly test ~50 different conditions in 1–2 h, which is ideal for testing biochemical fractions, titration of purified proteins or inhibitors, and time

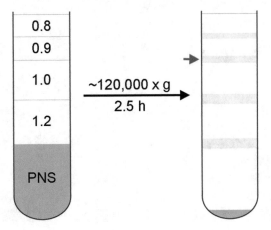

Fig. 2 Cartoon of sucrose step gradient before and after centrifugation. *Red arrow* indicates the 0.9/1.0 M sucrose interface where intact Golgi complexes are highly enriched

84   Edward B. Cluett et al.

course experiments. We have used it to demonstrate the requirement for cytoplasmic PLA$_2$ enzymes, including PAFAH Ib subunits [8, 28].

1. Quickly thaw out (e.g. in a 37 °C water bath) all solutions and reagents and put on ice.

2. Remove any protein aggregates from BBC and recombinant proteins of interest by ultracentrifugation (e.g. spin BBC in Beckman TLA 100.3 rotor at $264,600 \times g$ max for 20 min at 4 °C).

3. Place formvar/carbon-coated grids in self-closing forceps (Fig. 3a).

   - The forceps should hold grids on their very edge.
   - Use a white piece of paper as a background.
   - Check by eye to ensure grids are uniformly coated.

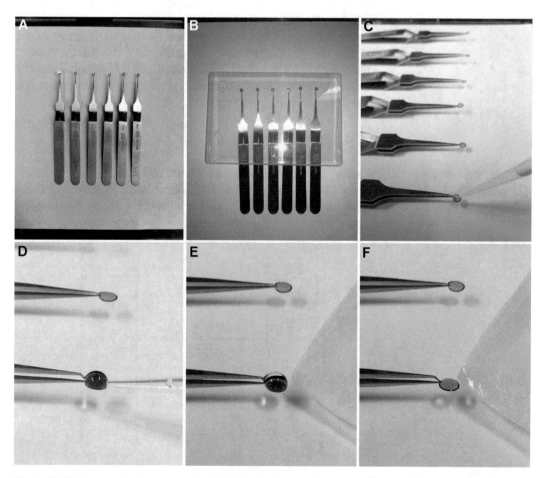

Fig. 3 (**a**) Hold formvar/carbon-coated grids with self-closing forceps. (**b**) Use a pipet tip box top (or similar cover) to protect grids from dust. (**c, d**) Gently deliver samples to the grids. (**e, f**) Blot off fluids with a moist Kimwipes

4. Place forceps on white paper and line them up with coated side up.

   – use the top of a transparent pipet tip box (or similar) to cover grids to prevent dust from settling (Fig. 3b).

5. Place the appropriate number of 500 μl microfuge tubes in the top of large pipet tip holder and label. Two per condition would be appropriate for duplicate samples.

6. Make up the ATP/Mg solution and put on ice: 80 μl reaction buffer, 10 μl ATP stock, 10 μl MgCl$_2$ stock.

7. Prepare the Reaction Mix by adding the dialysis buffer and ATP/MgCl$_2$ solutions together with source of tubulation activity (e.g., BBC) in a 500 μl microfuge tube. You need 20 μl/sample. Make enough to have an excess amount. The activity of the tubulation source (e.g., BBC or PLA$_2$ enzyme) should be determined by titration to get a dose-dependency curve. The proportions of reaction buffer and source of tubulation activity can vary depending on the amount of activity and protein in the unknown samples. The proportions can also vary when doing a titration experiment (*see* **Note 6**).

   – for solutions with high protein concentrations, e.g., BBC, start with the following proportions: 85 μl reaction buffer, 12 μl BBC, 1 μl ATP/Mg solution, 1 μl HEPES buffer.

   – for lower protein concentration material, e.g., protein fractions from columns: 98 μl fraction material, 1 μl ATP/Mg solution, 1 μl HEPES buffer.

   – for purified proteins: reaction buffer + purified protein(s) = 98 μl, 1 μl ATP/Mg solution, 1 μl HEPES buffer.

   – for negative control samples (i.e., without tubulation activity): 0.25 mg/ml of an inert/nonspecific protein, e.g., bovine serum albumin, or a non-tubulating PLA2 (snake venom PLA$_2$) should be used to assure even, negatively stained grids, otherwise samples will have non-uniform, splotchy staining (*see* **Note 7**).

   – add the above components in the order they are listed.

   – warm in 37 °C water bath for 5–15 min.

8. While the solution is warming, add 10 μl of Golgi suspension to each labeled microfuge tube.

9. Add 10 μl of Reaction Mix to each of the Golgi tubes (stagger addition of each sample by 1 min).

   – Add very carefully to the Golgi suspension and gently mix by inversion or gentle flicking once or twice after each drop (do not mix by pipetting). Make sure there are no Schlieren lines before incubating.

10. Incubate in 37 °C water bath for 0–15 min.

11. Take 10 μl of each sample and spot onto an EM grid (Fig. 3c, d). Let sample sit on grid for 15 min at room temperature. Cover as in **step 4**.

12. Carefully add 10 μl of 2 % PTA and let stand for 5 s.

13. Blot off/wick away the drop from the grid with damp Kimwipes (Fig. 3e, f).

14. Add 10 μl 2 % PTA two times, blotting off each time.

15. Let grids dry and then store in an appropriate grid storage container out of light until observation.

16. Image with a transmission EM (Fig. 4a, b) (*see* **Note 8**).

3.4 Immunogold Labeling of Negatively Stained Golgi Membranes

To characterize the biogenesis of Golgi membrane tubules, it might be useful to immunolocalize proteins of interest to the membranes, e.g., a known marker of Golgi membranes or a putative tubulation regulatory factor [27]. Golgi complexes can immunolabeled using a whole-mount, negative stain EM-immunogold protocol. Golgi-enriched fractions from rat liver (or other sources) are incubated in the presence of various concentrations of assorted combinations of buffer, BBC, and PLA$_2$ inhibitors to induce or suppress membrane tubulation, as described above. These in vitro samples are then treated as follows.

1. Place 10 μl drops of reaction mixtures on carbon- and Formvar-coated nickel grids for 15 min to allow attachment of Golgi complexes. Hold the grids with self-closing forceps as in Subheading 3.3, **step 2**.

2. Grids are then placed upside down on 0.5 ml of periodate-lysine-paraformaldehyde fixative for 5 min at room temperature (RT) in a closed microfuge tube.

3. Wash grids three times (10 min each) in buffer A at RT by transfer of the grids to a tube containing this solution.

4. Permeabilize Golgi membranes in Buffer B for 2 min at RT.

5. Repeat **step 3**.

6. Incubate grids in microfuge tubes containing 50 μl of primary antibody diluted in buffer A overnight at 4 °C. The appropriate concentration of the antibody solution will have to be empirically determined. Under these conditions, we have successfully used rabbit polyclonal anti-mannosidase II antibodies diluted 1:100 in buffer A.

7. Repeat **step 3**.

8. Repeat **step 2**.

9. Repeat **step 3**.

10. Incubate grids in microfuge tubes containing 50 μl of secondary antibody, e.g., goat anti-rabbit immunoglobulin G conjugated to 15-nm gold particles, diluted in buffer A for 3 h at RT.

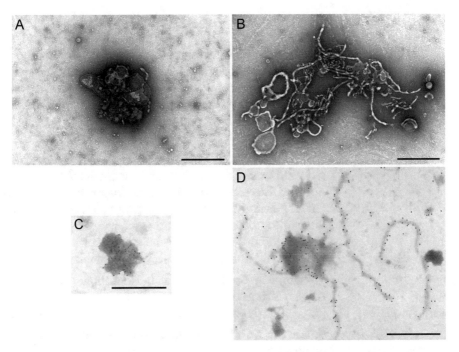

Fig. 4 Electron microscopy of negatively stained Golgi membranes and immunogold labeling of a resident Golgi membrane protein. (**a, b**) Golgi-enriched fractions were incubated in vitro and applied as whole-mount preparations onto EM grids for standard negative staining (**a, b**) or with a modified protocol involving a combination of immunogold labeling with anti-mannosidase I (ManII) antibodies followed by negative staining (**c, d**). (**a**) Golgi complexes incubated with buffer control show few membrane tubules. (**b**) Golgi complexes incubated with bovine brain cytosol (BBC) under conditions that induce membrane tubule formation. (**c**) Immunogold labeling with anti-ManII antibodies of control Golgi complexes incubated with buffer alone. (**d**) Golgi complexes incubated with BBC under tubulation conditions and then immunolabeled to show ManII distribution. Panels **c** and **d** were originally published in ref. [25]. Bar = 0.5 μm

11. Repeat **steps 7–9**.

12. Perform the negative staining procedure as in Subheading 3.3, **steps 13** and **14**.

13. Analyze by transmission EM (Fig. 4c, d).

4   Notes

1. The diameter of the ball bearing is dependent on the bore diameter of the homogenizer. The bore diameter of our homogenizer is different than the one published by Balch and Rothman [37]. The 0.2460 in. ball bearing we used allows a chamber clearance of 0.0054 in., with the clearance distance being the key. Pushing the minced liver through the syringes and homogenizer on the first pass takes a bit of effort. If the liver settles at the bottom of the syringe, the liquid/solid ratio

will be very different, and therefore it will be difficult to push the minced liver through the homogenizer. This can be remedied by inverting the homogenizer once with attached syringes before starting the passages. This will more evenly distribute the minced liver and ensure a more homogeneous, and therefore easier and more effective, homogenization. Inadequately minced liver will be very difficult or impossible to push through. Subsequent passes will be easier as the cells homogenize.

2. Typically male CD (Sprague-Dawley) rats are used, but female will yield similar results. Preparations using other rat strains were not as satisfactory. Our initial experiments used 150–200 g rats fed ad libitum. Livers obtained from older, larger rats are more difficult to homogenize. Young rats, weighing ~150–200 g, will yield livers and weighing ~7–10 g, which are ideal for homogenization and isolation of intact Golgi complexes.

3. We generally used calf brains because they were easier to process than those from adult animals.

4. If a peristaltic pump is not available, the saturated ammonium sulfate can be added by hand using partial volumes in 10 min increments.

5. There are multiple ways to prepare sucrose step gradients. The important point is to carefully add each sucrose step without disrupting the one underneath. We generally use an automatic pipettor, but this requires a deft touch and some practice. A safer but slower method is to use a Pasteur pipet.

6. We have used this assay to discover the role of PLA_2 enzymes in membrane tubule formation by incubating BBC or fractions thereof with PLA_2 antagonists. There are many PLA_2 antagonists, some with specificity for different classes of enzymes [40]. Also, some are reversible and some are irreversible, suicide substrates. Generally, each antagonist is pre-incubated with the Reaction Mix for a time sufficient to inhibit the enzyme(s). This pre-incubation may varying depending on the antagonist, and will likely have to be empirically determined. Also, it will important to perform the appropriate solvent controls. Sub-threshold BBC levels (i.e. amounts of BBC that yield background/baseline amounts of membrane tubules) can be used to examine activators of PLA_2-dependent tubule formation [41].

7. Isolated, intact Golgi complexes will exhibit varying amounts of membrane tubules. Therefore, it is important to always determine this baseline amount of tubulation for isolated Golgi complexes without tubule-stimulating components (e.g., without any BBC).

8. Under control conditions, intact Golgi complexes from rat liver appear as rounded piles of membranes with associated

vesicles but very few membrane tubules. When incubated with BBC or other preparations that induce membrane tubules, a variable number of membrane tubules or varying lengths will be generated (*see* Fig. 4).

Acknowledgement

This work was funded by NIH grant GM101027 to W.J.B.

References

1. Bonifacino JS, Glick BS (2004) The mechanisms of vesicle budding and fusion. Cell 116:153–166

2. De Matteis MA, Luini A (2008) Exiting the Golgi complex. Nat Rev Mol Cell Biol 9:273–284

3. Paczkowski JE, Richardson BC, Fromme JC (2015) Cargo adaptors: structures illuminate mechanisms regulating vesicle biogenesis. Trends Cell Biol 25:408–416

4. Bechler ME, de Figueiredo P, Brown WJ (2012) A PLA1-2 punch regulates the Golgi complex. Trends Cell Biol 22:116–124

5. Ha KD, Clarke BA, Brown WJ (2012) Regulation of the Golgi complex by phospholipid remodeling enzymes. Biochim Biophys Acta 1821:1078–1088

6. Polishchuk RS, Polishchuk EV, Marra P, Alberti S, Buccione R, Luini A, Mironov AA (2000) Correlative light-electron microscopy reveals the tubular-saccular ultrastructure of carriers operating between Golgi apparatus and plasma membrane. J Cell Biol 148:45–58

7. de Figueiredo P, Drecktrah D, Katzenellenbogen JA, Strang M, Brown WJ (1998) Evidence that phospholipase A$_2$ activity is required for Golgi complex and trans Golgi network membrane tubulation. Proc Natl Acad Sci U S A 95:8642–8647

8. de Figueiredo P, Polizotto RS, Drecktrah D, Brown WJ (1999) Membrane tubule-mediated reassembly and maintenance of the Golgi complex is disrupted by phospholipase A$_2$ antagonists. Mol Biol Cell 10:1763–1782

9. Yang JS, Valente C, Polishchuk RS, Turacchio G, Layre E, Moody DB, Leslie CC, Gelb MH, Brown WJ, Corda D, Luini A, Hsu VW (2011) COPI acts in both vesicular and tubular transport. Nat Cell Biol 13:996–1003

10. Lippincott-Schwartz J, Donaldson JG, Schweizer A, Berger EG, Hauri HP, Yuan LC, Klausner RD (1990) Microtubule-dependent retrograde transport of proteins into the ER in the presence of brefeldin A suggests an ER recycling pathway. Cell 60:821–836

11. Lippincott-Schwartz J, Yuan L, Tipper C, Amherdt M, Orci L, Klausner RD (1991) Brefeldin A's effects on endosomes, lysosomes, and the TGN suggest a general mechanism for regulating organelle structure and membrane traffic. Cell 67:601–616

12. Lippincott-Schwartz J, Yuan LC, Bonifacino JS, Klausner RD (1989) Rapid redistribution of Golgi proteins into the ER in cells treated with brefeldin A: evidence for membrane cycling from Golgi to ER. Cell 56:801–813

13. Wood SA, Park JE, Brown WJ (1991) Brefeldin A causes a microtubule-mediated fusion of the trans-Golgi network and early endosomes. Cell 67:591–600

14. Wood SA, Brown WJ (1992) The morphology but not the function of endosomes and lysosomes is altered by brefeldin-A. J Cell Biol 119:273–285

15. Jackson CL (2009) Mechanisms of transport through the Golgi complex. J Cell Sci 122:443–452

16. Jackson CL, Casanova JE (2000) Turning on ARF: the Sec7 family of guanine-nucleotide-exchange factors. Trends Cell Biol 10:60–67

17. Mollenhauer HH, Morre DJ (1966) Tubular connections between dictyosomes and forming secretory vesicles in plant Golgi apparatus. J Cell Biol 29:373–376

18. Mollenhauer HH, Morre DJ (1994) Structure of Golgi apparatus. Protoplasma 180:14–28

19. Marsh BJ, Mastronarde DN, Buttle KF, Howell KE, McIntosh JR (2001) Organellar relationships in the Golgi region of the pancreatic beta cell line, HIT-T15, visualized by high resolution electron tomography. Proc Natl Acad Sci U S A 98:2399–2406

20. Marsh BJ, Mastronarde DN, McIntosh JR, Howell KE (2001) Structural evidence for multiple transport mechanisms through the

Golgi in the pancreatic beta-cell line, HIT-T15. Biochem Soc Trans 29:461–467

21. Marsh BJ, Volkmann N, McIntosh JR, Howell KE (2004) Direct continuities between cisternae at different levels of the Golgi complex in glucose-stimulated mouse islet beta cells. Proc Natl Acad Sci U S A 101:5565–5570

22. Lippincott-Schwartz J, Cole N, Presley J (1998) Unravelling Golgi membrane traffic with green fluorescent protein chimeras. Trends Cell Biol 8:16–20

23. Lippincott-Schwartz J, Roberts TH, Hirschberg K (2000) Secretory protein trafficking and organelle dynamics in living cells. Annu Rev Cell Dev Biol 16:557–589

24. Banta M, Polizotto RS, Wood SA, de Figueiredo P, Brown WJ (1995) Characterization of a cytosolic activity that induces the formation of Golgi membrane tubules in a cell-free reconstitution system. Biochemistry 34:13359–13366

25. Cluett EB, Brown WJ (1992) Adhesion of Golgi cisternae by proteinaceous interactions - intercisternal bridges as putative adhesive structures. J Cell Sci 103:773–784

26. Cluett EB, Wood SA, Banta M, Brown WJ (1993) Tubulation of Golgi membranes in vivo and in vitro in the absence of Brefeldin-A. J Cell Biol 120:15–24

27. de Figueiredo P, Drecktrah D, Polizotto RS, Cole NB, Lippincott-Schwartz J, Brown WJ (2000) Phospholipase A₂ antagonists inhibit constitutive retrograde membrane traffic to the endoplasmic reticulum. Traffic 1:504–511

28. Bechler ME, Doody AM, Racoosin E, Lin L, Lee KH, Brown WJ (2010) The phospholipase complex PAFAH Ib regulates the functional organization of the Golgi complex. J Cell Biol 190:45–53

29. San Pietro E, Capestrano M, Polishchuk EV, DiPentima A, Trucco A, Zizza P, Mariggio S, Pulvirenti T, Sallese M, Tete S, Mironov AA, Leslie CC, Corda D, Luini A, Polishchuk RS (2009) Group IV phospholipase A₂α controls the formation of inter-cisternal continuities involved in intra-golgi transport. PLoS Biol 7:e1000194

30. Ben-Tekaya H, Kahn RA, Hauri HP (2010) ADP ribosylation factors 1 and 4 and group VIA phospholipase A regulate morphology and intraorganellar traffic in the endoplasmic reticulum-Golgi intermediate compartment. Mol Biol Cell 21:4130–4140

31. Bechler ME, Doody AM, Ha KD, Judson BL, Chen I, Brown WJ (2011) The phospholipase A₂ enzyme complex PAFAH Ib mediates endosomal membrane tubule formation and trafficking. Mol Biol Cell 22(13):2348–2359

32. Capestrano M, Mariggio S, Perinetti G, Egorova AV, Iacobacci S, Santoro M, Di Pentima A, Iurisci C, Egorov MV, Di Tullio G, Buccione R, Luini A, Polishchuk RS (2014) Cytosolic phospholipase A₂ε drives recycling through the clathrin-independent endocytic route. J Cell Sci 127:977–993

33. de Figueiredo P, Doody A, Polizotto RS, Drecktrah D, Wood S, Banta M, Strang M, Brown WJ (2001) Inhibition of transferrin recycling and endosome tubulation by phospholipase A₂ antagonists. J Biol Chem 276:47361–47370

34. Doody AM, Antosh AL, Brown WJ (2009) Cytoplasmic phospholipase A₂ antagonists inhibit multiple endocytic membrane trafficking pathways. Biochem Biophys Res Commun 388:695–699

35. Bankaitis VA (2009) The Cirque du Soleil of Golgi membrane dynamics. J Cell Biol 186:169–171

36. Schmidt JA, Brown WJ (2009) Lysophosphatidic acid acyltransferase 3 regulates Golgi complex structure and function. J Cell Biol 186:211–218

37. Balch WE, Rothman JE (1985) Characterization of protein transport between successive compartments of the Golgi apparatus: asymmetric properties of donor and acceptor activities in a cell-free system. Arch Biochem Biophys 240:413–425

38. McLean IW, Nakane PK (1974) Periodate-lysine-paraformaldehyde fixative. A new fixation for immunoelectron microscopy. J Histochem Cytochem 22:1077–1083

39. Malhotra V, Serafini T, Orci L, Shepherd JC, Rothman JE (1989) Purification of a novel class of coated vesicles mediating biosynthetic protein transport through the Golgi stack. Cell 58:329–336

40. Brown WJ, Chambers K, Doody A (2003) Phospholipase A₂ (PLA₂) enzymes in membrane trafficking: mediators of membrane shape and function. Traffic 4:214–221

41. Polizotto RS, de Figueiredo P, Brown WJ (1999) Stimulation of Golgi membrane tubulation and retrograde trafficking to the ER by phospholipase A₂ activating protein (PLAP) peptide. J Cell Biochem 74:670–683

42. Boncompain G, Divoux S, Gareil N, de Forges H, Luscure A, Latrechi L, Mercanti V, Jollivet F, Raposo G, Perez F (2012) Synchronization of secretory protein traffic in populations of cells. Nat Methods 9:493–498

Chapter 8

Proteomic Characterization of Golgi Membranes Enriched from Arabidopsis Suspension Cell Cultures

Sara Fasmer Hansen, Berit Ebert, Carsten Rautengarten, and Joshua L. Heazlewood

Abstract

The plant Golgi apparatus has a central role in the secretory pathway and is the principal site within the cell for the assembly and processing of macromolecules. The stacked membrane structure of the Golgi apparatus along with its interactions with the cytoskeleton and endoplasmic reticulum has historically made the isolation and purification of this organelle difficult. Density centrifugation has typically been used to enrich Golgi membranes from plant microsomal preparations, and aside from minor adaptations, the approach is still widely employed. Here we outline the enrichment of Golgi membranes from an Arabidopsis cell suspension culture that can be used to investigate the proteome of this organelle. We also provide a useful workflow for the examination of proteomic data as the result of multiple analyses. Finally, we highlight a simple technique to validate the subcellular localization of proteins by fluorescent tags after their identification by tandem mass spectrometry.

Key words Golgi apparatus, Density gradient centrifugation, Subcellular localization, Fluorescent protein

1 Introduction

The Golgi apparatus is a unique and complex structure within the eukaryotic cell. The organelle is composed of flat membranes called cisternae that interlink to form stacks [1]. The plant Golgi apparatus is a major junction of the secretory system where proteins, lipids, carbohydrates are processed and biosynthesized prior to their distribution to all parts of the cell. This includes the synthesis of matrix polysaccharides destined for the cell wall [2, 3], the production of complex N-glycans [4], the synthesis of glycolipids [5], sequestration of protein complexes [6], intracellular signaling [7], and trafficking of macromolecules [8].

The Golgi apparatus is an integral component of the plant endomembrane [9] and along with the endoplasmic reticulum (ER) is one of the more difficult structures to purify from plant tissue [10]. In fact it is this very proximity of the Golgi to the ER [11] and the

William J. Brown (ed.), *The Golgi Complex: Methods and Protocols*, Methods in Molecular Biology, vol. 1496,
DOI 10.1007/978-1-4939-6463-5_8, © Springer Science+Business Media New York 2016

intimate interactions with the cytoskeleton [12] that has historically made the enrichment and purification of this structure problematic. These difficulties in purifying Golgi bodies from plant material has resulted in the development of more advanced purification techniques such as LOPIT [13, 14] and the application of free-flow electrophoresis [15–17]. Nonetheless, it is still suitable to enrich Golgi membranes from plant material for a variety of downstream applications, including proteomic surveys, using traditional density centrifugation procedures. In fact, the utilization of density centrifugation is still the most widely used approach for the enrichment of Golgi membranes from microbes [18], animals [19], and plants [20, 21], although it should be noted that Golgi membrane fractions isolated from plant material by density centrifugation alone are usually not of sufficient purity to conduct a thorough proteomic characterization of this organelle [22]. However, the recent high purity surveys conducted on the Arabidopsis Golgi proteome [14, 16] in conjunction with our functional knowledge of this organelle means that a comparative or quantitative survey conducted on enriched material should still be considered a viable approach.

Here we outline a detailed protocol for the enrichment of Golgi membranes from an Arabidopsis suspension cell culture using a discontinuous sucrose gradient that would be suitable for proteomic investigations. A proteomic data analysis workflow is also provided to assist in the identification of high confidence candidates after mass spectrometry. Finally, an approach employing transient expression in *Nicotiana benthamiana* is outlined for the subcellular validation of candidate proteins by fluorescent tags is also detailed.

2 Materials

Prepare solutions with ultrapure water (18 MΩ cm at 25 °C) and analytical grade reagents. Utilize higher grade reagents (LC-MS grade) for solutions and buffers used in conjunction with mass spectrometry. Prepare reagents at room temperature. Unless otherwise stated, prepare buffers the day before and store at 4 °C. Perform all centrifugation steps at 4 °C.

2.1 Arabidopsis Suspension Cell Cultures

1. Arabidopsis suspension cell cultures (*see* **Note 1**).
2. Temperature controlled shaking incubator (22 °C, 120 rpm) with constant light (100 μE).
3. Arabidopsis Cell Culture Medium: 2% (w/v) sucrose, α-naphthaleneacetic acid (0.5 mg/L), kinetin (0.05 mg/L), 1× Murashige and Skoog basal salt mixture [23]. Prepare media and adjust to pH 5.7 with potassium hydroxide (KOH), autoclave for 20 min at 121 °C and store at 4 °C (*see* **Note 2**).
4. 250 mL glass Erlenmeyer flasks.

2.2 Protoplast Preparation of Arabidopsis Suspension Cell Cultures

1. Digestion buffer: 500 mM mannitol, 5 mM 2-(N-morpholino) ethanesulfonic acid (MES), adjust to pH 5.7 with KOH. Store at 4 °C. Just prior to use, add 0.4 % (w/v) cellulase "Onozuka" R-10 and 0.05 % (w/v) pectolyase Y-23 (Yakult Pharmaceutical).

2. Protoplast wash buffer: 500 mM Mannitol, 5 mM 2-(N-morpholino) ethanesulfonic acid (MES), adjust to pH 5.7 with KOH. Store at 4 °C.

3. Miracloth (Merck Millipore).

4. Variable speed benchtop orbital shaker (*see* **Note 3**).

5. Preparative centrifuge for 250 mL tubes and capacity to $800 \times g$, such as an Avanti J25 centrifuge (Beckman Coulter) with a JLA-16.250 rotor (Beckman Coulter).

2.3 Protoplast Homogenization

1. Homogenization buffer: 1% (w/v) dextran (Mw 200,000), 0.4 M sucrose, 10 mM disodium hydrogen phosphate (Na_2HPO_4), 3 mM ethylenediaminetetraacetic acid (EDTA), 0.1 % (w/v) bovine serum albumin (BSA), 5 mM dithiothreitol (DTT) (*see* **Note 4**), pH to 7.1 with sodium hydroxide (NaOH).

2. Glass-Teflon Potter-Elvehjem Tissue Grinder with a 20–40 mL capacity (*see* **Note 5**).

3. Preparative centrifuge for 50 mL tubes and capacity to $5000 \times g$, such as an Avanti J25 centrifuge (Beckman Coulter) with a JA-25.50 rotor (Beckman Coulter).

4. Light microscope capable of visualizing plant cells and large subcellular structures, minimum 40× objective.

2.4 Enrichment of Golgi Membranes by Differential Centrifugation

1. Gradient buffer 1: 1.4 M sucrose, 10 mM Na_2HPO_4, 3 mM EDTA, pH 7.1 with NaOH, can store at –20 °C.

2. Gradient buffer 2: 1.0 M sucrose, 10 mM Na_2HPO_4, 3 mM EDTA, dextran Mw 200,000 (1 % w/v), 5 mM DTT (*see* **Note 4**), pH 7.1 with NaOH, can store at –20 °C.

3. Gradient buffer 3: 0.2 M sucrose, 10 mM Na_2HPO_4, 3 mM EDTA, dextran Mw 200,000 (1 % w/v), 5 mM DTT (*see* **Note 4**), pH 7.1 with NaOH, can store at –20 °C.

4. Gradient buffer 4: 0.1 M sucrose, 10 mM Na_2HPO_4, 3 mM EDTA, dextran Mw 200,000 (1 % w/v), 5 mM DTT (*see* **Note 4**), pH 7.1 with NaOH, can store at –20 °C.

5. Ultracentrifuge with swing-out rotor for 40 mL tubes and capable of $100,000 \times g$ for gradients, such as an Optima™ XE (Beckman Coulter) with a SW 30 Ti rotor (Beckman Coulter).

6. Disposable plastic pipettes (1 mL).

7. 5 M KCl.

8. 10 mM Tris–HCl, pH 8.5.

9. Ultracentrifuge with fixed angle rotor for 2 mL tubes and capable of 100,000×*g*, such as an Optima™ MAX-TL (Beckman Coulter) with a TLA-100 rotor (Beckman Coulter).

10. Protein Quantification Assay, such as Pierce™ BCA Protein Assay Kit.

2.5 Analysis of Golgi Enriched Membranes by Mass Spectrometry (LC-MS/MS)

1. Tandem mass spectrometer (MS/MS) with liquid chromatography (LC) delivery system capable of data dependent acquisition (DDA)/independent data acquisition (IDA) such as the Q Exactive™ Hybrid Quadrupole-Orbitrap Mass Spectrometer (Thermo Scientific) with a Proxeon Easy-nLC II HPLC (Thermo Scientific) (*see* **Note 6**).

2. Digestion buffer: 1 M urea and 10 mM tris(hydroxymethyl) aminomethane (Tris–HCl), pH 8.5 solution (*see* **Note 7**).

3. 1 M dithiothreitol (DTT) (*see* **Note 8**).

4. 1 M iodoacetamide (IAA) (*see* **Note 9**).

5. High-grade trypsin, such as trypsin from porcine pancreas (Sigma-Aldrich).

6. Solid phase extraction (SPE) for peptides, such as Micro SpinColumns with C_{18} (Harvard Apparatus, MA, USA).

7. SPE buffer 1: 80% acetonitrile (v/v) with 0.1% trifluoroacetic acid (v/v).

8. SPE buffer 2: 2% acetonitrile (v/v) with 0.1% trifluoroacetic acid (v/v).

9. MS buffer A: 2% acetonitrile, 0.1% formic acid.

10. SpeedVac concentrator.

11. Access to search engine to identify proteins from tandem mass spectrometry data, such as Mascot (Matrix Science) (*see* **Note 10**).

12. Proteome integration, profiling and quantitation software, such as Scaffold 3 (Proteome Software) (*see* **Note 11**).

2.6 Validation of Golgi Localization by Fluorescent Protein Fusions

2.6.1 Cultivation of Nicotiana benthamiana

1. Plastic plant pots (80×80 mm).

2. Germination tray (280×540 mm) with a transparent plastic lid.

3. *Nicotiana benthamiana* "Domin" seeds.

4. Soil, such as PRO-MIX (Premier Horticulture).

5. Plant growth chamber (*see* **Note 12**).

2.6.2 Plasmid Preparation and Agrobacterium tumefaciens Transformation

1. cDNA from plant material (*see* **Note 13**).

2. Thermocycler (PCR machine).

3. Gene specific primers (*see* **Note 14**).

4. DNA Polymerase, e.g., Phusion™ High-Fidelity DNA Polymerase (Thermo Scientific), dNTPs (*see* **Note 15**).

5. Agarose gel electrophoresis equipment.

6. Agarose, e.g., UltraPure™ Agarose (Life Technologies).

7. 1× TAE buffer, e.g., TAE Buffer (Tris–acetate–EDTA) (50×).

8. DNA loading buffer, e.g., Gel Loading Solution.

9. DNA ladder, e.g., 1 Kb Plus DNA Ladder.

10. UV light box.

11. Gel and PCR cleanup kit, e.g., NucleoSpin® Gel and PCR cleanup.

12. Plasmid preparation kit, e.g., QIAprep Spin Miniprep Kit (Qiagen).

13. pCR®8/GW/TOPO® TA Cloning Kit.

14. Gateway® LR Clonase® II Enzyme mix.

15. Gateway compatible vector containing a fluorescent protein, e.g., pEarleyGate [24] (*see* **Note 16**).

16. Competent *Escherichia coli* (*E. coli*), e.g., One Shot® TOP10 Chemically Competent *E. coli*.

17. Spectinomycin, kanamycin, and gentamycin.

18. Temperature mixer, such as the Thermomixer compact.

19. Electro-competent *Agrobacterium*, e.g., strain GV3101::pMP90 (*see* **Note 17**).

20. Disposable electroporation cuvettes with 1 or 2 mm gap sizes.

21. Electroporation system, such as a Gene Pulser Xcell™ Electroporation Systems.

22. Luria–Bertani (LB) media: 10 g tryptone, 10 g NaCl, 5 g yeast extract, adjust to pH 7 with NaOH and sterilize by autoclaving.

23. Luria–Bertani (LB) media supplemented with agar: 10 g tryptone, 10 g NaCl, 5 g yeast extract, 7.5 g agar in 1 L water, adjust to pH 7 with NaOH and sterilize by autoclaving.

2.6.3 Infiltration

1. Luria–Bertani (LB) media: 10 g tryptone, 10 g NaCl, 5 g yeast extract, 7.5 g agar in 1 L water, adjust to pH 7 with NaOH and sterilize by autoclaving.

2. Sterile pipette tips or toothpicks.

3. Preparative centrifuge, such as Allegra 25R Benchtop Centrifuge.

4. Infiltration buffer: 10 mM 2-(*N*-morpholino)ethanesulfonic acid (MES) pH 5.6, 10 mM $MgCl_2$, 100 µM acetosyringone (*see* **Note 18**).

5. 1 mL disposable polypropylene syringes.

2.6.4 Confocal Microscopy

1. Confocal laser scanning microscope (*see* **Note 19**).

2. Coverslip holder, such as Attofluor® cell chamber for microscopy.

3. 25 mm round glass coverslips.

4. 10% glycerol.

3   Methods

3.1   Growth of Arabidopsis Suspension Cell Cultures

1. The Arabidopsis suspension cell cultures are cultivated in 120 mL aliquots.

2. To maintain healthy cultures, the cells are subcultured weekly into a flask containing 100 mL of Arabidopsis cell culture medium.

3. Cells are subcultured at a ratio of 1:5–1:10, depending on density of 7-day-old cultures (*see* **Note 20**).

4. Typically, 15–20 mL from a 7-day culture is added to a new flask containing 100 mL growth media.

3.2   Preparation of Protoplasts

1. Filter 7-day-old cells from approximately four to five flasks through Miracloth and squeeze out the remaining culture medium to yield around 50 g of cells fresh weight.

2. Add 250 mL of digestion buffer to a 500 mL beaker and then add 0.4% (w/v) cellulase and 0.05% (w/v) pectolyase and resuspend the enzymes (*see* **Note 21**). Add the 50 g of cells and gently mix, then gently transfer to a 4 L wide-bottomed conical flask (*see* **Note 22**).

3. Place on an orbital shaker and rotate slowly in the dark (wrapped in foil) for 3 h (*see* **Note 23**).

4. After 3 h, protoplasts are harvested by centrifugation at $800 \times g$ for 5 min.

5. Gently resuspended the cell pellet in about 100 mL of digestion buffer (no added enzymes), centrifuge at $800 \times g$ and discard the buffer. Repeat this step two more times to ensure complete removal of cellulase and pectolyase from the protoplasts (*see* **Note 24**).

6. After the final wash, discard the buffer and place the protoplasts on ice.

3.3   Disruption of Arabidopsis Protoplasts

1. Resuspend the pellet/protoplasts in homogenization buffer using a minimum ratio of 1:1 (w/v) and keep on ice. This volume is determined using the original fresh weight of the cells.

2. Transfer about 10–20 mL of protoplasts in the homogenization buffer into the Potter-Elvehjem homogenizer.

3. Disrupt the protoplasts using four to five strokes at even pressure of the pestle in the Potter-Elvehjem homogenizer (*see* **Note 25**). Transfer disrupted protoplasts to a 250 mL beaker on ice.

4. Prior to processing/rupturing the rest of the protoplasts, check to ensure disruption was effective by examining homogenate with a light microscope.

5. If over 75 % of cells have been ruptured, process the rest of the protoplast homogenate, otherwise process the initial sample again with further strokes in the Potter-Elvehjem homogenizer.

6. Centrifuge the homogenate at $5000 \times g$ for 15 min.

7. Carefully decant supernatant into a 250 mL beaker on ice.

3.4 Enrichment of Golgi Membranes

1. Add 5 mL of gradient buffer 1 to two 40 mL ultracentrifuge tubes. With a disposable plastic pipette, gently layer the homogenate on to the gradient buffer 1 until all tubes are at least two thirds full (Fig. 1).

2. Ultracentrifuge homogenate at $100,000 \times g$ for 1 h (*see* **Note 26**).

3. Remove the supernatant without disturbing the membrane rich lysate that has formed on the sucrose cushion (gradient buffer 1) (*see* **Note 27**).

4. Once the supernatant has been removed, carefully layer 15–20 mL of gradient buffer 2 on top of the membrane rich lysate. Follow this with 8–10 mL of gradient buffer 3 and then 5–8 mL of gradient buffer 4 (*see* **Note 28**).

5. Ultracentrifuge samples at $100,000 \times g$ for 90 min.

6. After ultracentrifugation, remove tubes and inspect the banding patter. Two distinct bands should be present above the 1.0 M gradient buffer (Fig. 1). Gently remove the upper band between the 0.1 and 0.2 M gradient buffers with a disposable plastic pipette and discard. Then carefully remove the remaining upper band containing the enriched Golgi membranes situated on top of the 1.0 M gradient buffer and place in a 15 mL tube on ice (*see* **Note 29**).

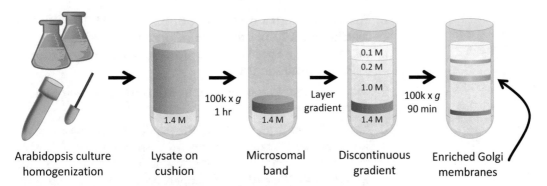

Fig. 1 Flowchart outlining the procedure for the enrichment of Golgi membranes from Arabidopsis suspension cell cultures using a discontinuous gradient

7. Estimate the protein concentration of the extract using a protein quantification assay and transfer the sample into 10 μg aliquots and place on ice.

8. Add 5 M KCl to achieve an estimated final concentration of about 50 mM to each 10 μg aliquot of enriched Golgi membrane proteins fractions and incubate for 15 min at 4 °C, gently shaken (see **Note 30**).

9. Ultracentrifuge samples at $100,000 \times g$ for 60 min and discard the supernatant.

10. Wash the pellet twice with 1 mL 10 mM Tris–HCl pH 8.5.

11. At this stage, any 10 μg salt washed Golgi membrane aliquots not being processed for analysis by tandem mass spectrometry can be stored at –80 °C.

3.5 Characterization of Enriched Golgi Fractions by Tandem Mass Spectrometry

1. Resuspend the 10 μg salt washed Golgi membrane pellet in about 50 μL of digestion buffer.

2. Add DTT to the diluted protein extract to a final concentration of 25 mM and incubate 30 min (room temperature).

3. Then add IAA to a final concentration of 50 mM and incubate 30 min (room temperature) in the dark.

4. Add trypsin at a 1:10 trypsin–protein ratio (see **Note 31**) and incubate overnight (37 °C) (see **Note 32**).

5. Remove urea and concentrate samples with a Micro SpinColumn (25–75 μL capacity). Initially hydrate the C_{18} matrix with ultrapure water (75 μL) for 10 min and centrifuge ($1000 \times g$, 2 min) as per manufacturer's instructions.

6. Wash the SpinColumn with the 50 μL SPE Buffer 1 and centrifuge ($1000 \times g$, 2 min) and prime twice with 50 μL SPE Buffer 2, centrifuging ($1000 \times g$, 2 min) after each step.

7. Add digested peptides in urea solution to the SpinColumn and centrifuge ($1000 \times g$, 2 min), wash twice with 50 μL SPE Buffer 2, centrifuging ($1000 \times g$, 2 min) after each step.

8. Finally elute into a new tube with 25–50 μL SPE Buffer 1 by centrifuging $1000 \times g$ for 2 min. Concentrate and remove acetonitrile with a SpeedVac concentrator until 1–5 μL of peptide solution remains in the tube.

9. Dilute around 4 μg of peptide in MS Buffer A to a volume of about 16 μL (see **Note 33**).

10. Analyze about 1 μg (4 μL) of the peptide sample by nanoflow liquid chromatography tandem mass spectrometry (LC-MS/MS) using an automated data dependent acquisition method optimized for the analysis of complex proteomes (see **Note 34**).

11. Data produced after LC-MS/MS analysis can be interrogated with search algorithms, such as the software package Mascot (Fig. 2a) to identify proteins found in each sample (see **Note 35**).

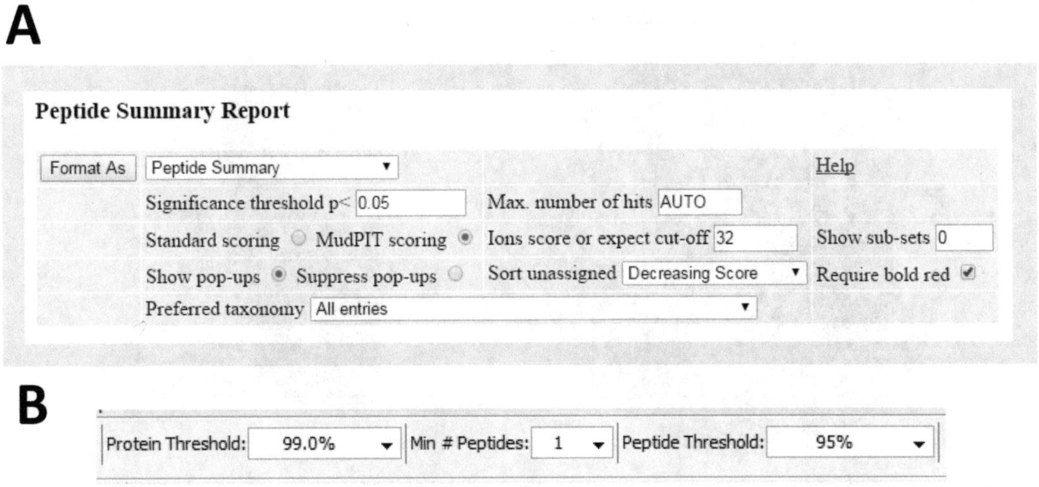

A

Peptide Summary Report

Format As | Peptide Summary ▼   Help

Significance threshold p< 0.05   Max. number of hits AUTO

Standard scoring ○ MudPIT scoring ⦿   Ions score or expect cut-off 32   Show sub-sets 0

Show pop-ups ⦿ Suppress pop-ups ○   Sort unassigned Decreasing Score ▼   Require bold red ☑

Preferred taxonomy All entries ▼

B

Protein Threshold: 99.0% ▼ | Min # Peptides: 1 ▼ | Peptide Threshold: 95% ▼

Fig. 2 Recommended protein and peptide filters for the analysis of mass spectrometry data derived from the analysis of Arabidopsis samples **(a)** Mascot (Matrix Science) and **(b)** Scaffold (Proteome Software)

12. The integration and relative quantitation of proteins identified in samples by LC-MS/MS can be achieved using software such as Scaffold, which can be used to merge multiple results from Mascot. The software enables further filters to be applied to both identified peptides and proteins (Fig. 2b), further ensuring only high confidence protein matches are identified (*see* **Note 36**).

13. When analyzing well-characterized proteomes from systems like Arabidopsis, resources can be readily employed to analyze results. Useful approaches involve functional categorizations using Gene Ontologies [25] or the subcellular distribution of a proteome using resources like the SUBcellular Arabidopsis database [26] (Fig. 3).

3.6 Transient Protein Expression for Subcellular Localization

To validate proteins identified in enriched Golgi membrane preparations it is possible to easily verify their location to this organelle using the following procedure.

3.6.1 Cultivation of Nicotiana Benthamiana

1. Plastic pots of around 80×80 mm containing high quality soil, such as PRO-MIX HP MYCORRHIZAE™ should be thoroughly watered and then drained.

2. About 30 seeds of *N. benthamiana* "Domin" are spread onto the moist soil, and pots put onto a plastic tray and covered with a transparent plastic lid.

3. Place the trays in a 25 °C chamber with 60% humidity and 16 h light–8 h dark cycle.

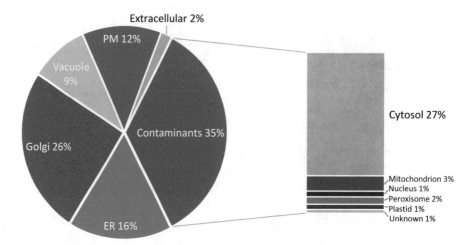

Fig. 3 Subcellular breakdown of a typical dataset resulting from a proteomic analysis of enriched of Golgi membranes from the Arabidopsis suspension cell culture. The analysis outlines proteins identified from two independent enrichment experiments using the outlined enrichment and data workflow. In total, 1102 proteins were identified after both samples were combined using Scaffold. The subcellular locations were obtained from the SUBcellular Arabidopsis (SUBA) database [29] and distributions are shown according to the total normalized spectral abundance factor (NSAF) [30], obtained from Scaffold, for proteins allocated to each compartment

4. After about a week, remove the plastic lid and propagate the plants under the same growth conditions for another week.

5. Transfer the 2-week-old seedlings into fresh pots (as previously described). Transplant one seedling per pot (*see* **Note 37**) and continue to propagate plants under identical growth conditions until ready to be infiltrated with *Agrobacterium* (*see* **Note 38**).

3.6.2 Plasmid Preparation and Transformation of Agrobacterium

1. Amplify the gene of interest by PCR employing gene specific primers and appropriate plant cDNA (*see* **Note 39**).

2. Analyze the resultant PCR product using a 1% agarose gel with 1× TAE buffer.

3. Visualize the PCR product using a UV light box and confirm the expected size.

4. Excise the DNA band and extract the DNA from the agarose gel using a gel cleanup kit.

5. Clone the purified PCR product with the pCR®8/GW/TOPO® TA Cloning Kit into the pCR®8 donor vector to create the entry clone.

6. Transform competent *E. coli* with the cloning reaction and select for positive transformant on LB plates with 100 μg/mL spectinomycin.

7. Purify the plasmid DNA from an overnight *E. coli* culture (LB with 100 μg/mL spectinomycin) using a Plasmid preparation kit and confirm the identity of the entry clone by sequencing.

8. Use the confirmed entry clone in a LR Clonase II Enzyme reaction and clone it into a destination vector (e.g., pEarley-Gate) containing an in-frame fluorescent protein.

9. Transform this reaction into *E. coli* and select on LB plate containing kanamycin (50 μg/mL). Using a sterile toothpick, transfer a colony to 5 mL LB with kanamycin (50 μg/mL) and grow at 37 °C overnight with shaking. Purify the plasmid using a plasmid preparation kit and confirm the integrity of the clone by sequencing.

10. Add 1 μL of the expression vector (e.g., pEarleyGate containing gene of interest) to competent *Agrobacterium* and sit on ice for 15 min (*see* **Note 40**).

11. Set up the electroporation device according to the manufacturer's instructions (*see* **Note 41**), transfer mixture to pre-chilled cuvette and apply voltage to transform *Agrobacterium*.

12. Immediately after the electroporation add 1 mL of LB media to the cuvette (*see* **Note 42**) and mix by gently pipetting.

13. Transfer the contents of the cuvette into a sterile 1.5 mL microfuge tube and incubate for at least 1 h at 30 °C with constant agitation.

14. Pipette about 50 μL (*see* **Note 43**) onto LB plates containing appropriate selective antibiotics (*see* **Note 44**) and incubate for 48 h at 30 °C until colonies are visible.

3.6.3 Infiltration of N. benthamiana

1. Inoculate 10 mL LB media (containing the appropriate selective antibiotics) with an *Agrobacterium* colony and grow overnight at 30 °C with constant shaking to log phase.

2. Centrifuge the overnight culture for 10 min at 4000×*g* at 20 °C.

3. Remove the supernatant, wash the pellet with 10 mL of infiltration buffer (*see* **Note 45**) and repeat the centrifugation step.

4. Resuspend the *Agrobacterium* in infiltration buffer to an optical density (OD) of 0.01–0.3 at 600 nm (*see* **Note 46**).

5. Infiltrate the abaxial surface of leaves from 3- to 4-week-old *N. benthamiana* plants using a 1 mL plastic syringe without a needle (*see* **Note 47**). Hold the adaxial side of the leaf firmly and press the syringe from abaxial side against your finger. Infiltrated areas will turn a darker green (*see* **Note 48**).

6. Grow infiltrated plants for an additional 2 days before examining expression of the fluorescent protein by confocal microscopy (*see* **Note 49**).

3.6.4 Confocal Microscopy

1. Use a sterile razor blade to cut a 12 × 12 mm piece from the infiltrated area of the leaf (*see* **Note 50**).

2. Place the leaf section onto a coverslip and add a drop of 10% glycerol on its surface.

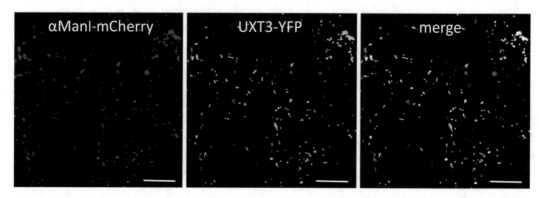

Fig. 4 The transient subcellular localization of the *cis*-Golgi marker α-Mannosidase I-mCherry (αManI-mCherry) and a resident Golgi protein (UXT3-YFP) [31] with an overlay (merge) indicting co-localization. Bar = 25 μm

3. Put a second coverslip on top and mount into a coverslip holder.

4. Place the slide onto the confocal stage and focus using bright-field at low magnification.

5. At this magnification use an appropriate filter or setting, depending on the fluorescent protein employed, to find cells exhibiting a florescent signal. Once a cell is located, a higher magnification can be employed.

6. Configure the confocal to enable sequential or simultaneous acquisitions (*see* **Note 51**).

7. Perform a live scan using the confocal. In our example, we employed a simultaneous acquisition for yellow fluorescent protein (YFP) and mCherry after co-infiltrating a vector containing UXT3-YFP and the Golgi marker, α-Mannosidase I-mCherry (Fig. 4).

8. Images can be processed with the microscope software such as ImageJ [27].

4   Notes

1. The Arabidopsis suspension cell cultures outlined here were originally created over 20 years ago from stem explants of the accession Landsberg erecta [28]. While any suspension cell culture should produce similar results, in our hands, this particular suspension culture contains a population of Golgi membranes which are minimally interconnected with the cytoskeleton. This results in enriched Golgi membranes with reduced contamination when compared to Arabidopsis seedlings or recently created cell lines. This particular cell line has been distributed globally and can be sourced through diligent inquiries.

2. Prior to autoclaving, media can be dispensed into 250 mL glass Erlenmeyer flasks in 100 mL aliquots and sealed with foil until required. Large batches can be premade (50–100) and stored at 4 °C in the dark.

3. A variable speed benchtop orbital shaker with a relatively large orbital throw is optimal (at least 2 cm). This enables a slow rotation while maintaining cells in solution and provides efficient enzymatic digestion of the cell wall.

4. The DTT should be added immediately prior to use. A stock solution of DTT (1 M) can be stored in aliquots at –20 °C if required.

5. The performance of the homogenizer is very important to gently rupturing protoplasts. Generally, negative pressure on the up-stroke should result in an air bubble between the Teflon pestle and the homogenate. This seems to be indicative of adequate restriction between the Teflon pestle and the glass mortar which constricts the protoplasts as they pass, resulting in their disruption. The performance of the homogenizer (generation of the air bubble) can be tested with water.

6. While access to high resolution tandem mass spectrometry has become somewhat common, they are expensive instruments to purchase and maintain. As a consequence many institutions provide access via dedicated facilities. While these facilities can seem like an expensive option, current pricing makes them highly completive when analyzing up to 50 samples per year.

7. Urea is unstable and can degrade in solution when exposed to elevated temperatures (>25 °C) or over time. This degradation results in the generation of isocyanic acid which subsequently reacts with the amino terminus of proteins as well as the sidechains of lysine and arginine residues. Consequently, it is recommended that urea solutions be made fresh as required. A 2 mL urea/Tris–HCl solution is relatively simple to make and usually adequate for most purposes.

8. A stock solution of DTT can be stored at –20 °C in aliquots and thawed as required.

9. A stock solution of IAA can be stored at –20 °C in aliquots and thawed as required. IAA is used to alkylate the thiol group on cysteine residues after reduction with DTT. It is virtually impossible to detect cysteine containing peptides unless controlled alkylation is undertaken. However, given the low occurrence of cysteine residues in proteins, the absence of this step has little impact on the final results, namely number of proteins identified and sequence coverage. IAA is light and heat sensitive and should be stored in the dark or wrapped in foil.

10. Suppliers of most tandem mass spectrometers designed for standard proteomic workflows will usually supply software

with their instruments that can be used for the interrogation of tandem spectra for the identification of proteins. There are also a number of open source software packages available as well as a range of third party software that can also be employed to identify proteins from fragmentation spectra. In general, these algorithms are relatively similar, we generally use Mascot (a third party software package from Matrix Science) and have found that it provides quality results when high confidence cutoffs are employed. Access to the software is free if submitting less than 1200 spectra in a single submission (http://www.matrixscience.com/).

11. A considerable data handling problem when dealing with results from multiple analyses of tandem spectra is data integration. While some software platforms have developed workflows to integrate multiple samples, Scaffold from Proteome Software (http://www.proteomesoftware.com/) provides a powerful way to combine and compare protein identifications from multiple samples. The software provides some quantitation support via spectral counting to enable limited comparative assessment between samples.

12. A growth chamber capable of maintaining 25 °C and 60% humidity with a day/night cycle is optimal. However, plants can also be grown in a glasshouse with some temperature control.

13. cDNA can be obtained from the plant tissue of interest using a plant RNA extraction kit, such as the RNeasy Plant Mini Kit (Qiagen). The mRNA is then reverse transcribed into cDNA using a Reverse Transcriptase, such as Superscript III Reverse Transcriptase (Life Technologies).

14. Ensure the reviewer primer does not contain a stop codon to enable a C-terminal fusion with the fluorescent protein.

15. To amplify a gene of interest and avoid errors, a proofreading DNA polymerase should be used.

16. The pEarleyGate vector collection generally works well; however, many other vectors are available that contain strong promoters and fluorescent tags.

17. *Agrobacterium tumefaciens* strain GV3101::pMP90 is a commonly used and works well when transforming *N. benthamiana* and can also be used to stably transform *Arabidopsis thaliana*. However, other *Agrobacterium* strains are available that also work in *N. benthamiana*, including C58C1, EHA105, LBA4404, and AGL1.

18. Prepare the infiltration buffer fresh for each infiltration. Stock solutions of 0.5 M MES-KOH pH 5.5 and 1 M $MgCl_2$ can be prepared in advance but should be autoclaved. For the 0.5 M MES stock solution, prepare MES first and adjust the pH to 5.5 with KOH. Acetosyringone should be added separately to

the infiltration buffer on the day of use. Prepare a 100 mM acetosyringone stock solution in either DMSO or 96 % (v/v) ethanol, and store in aliquots at –20 °C or for shorter periods at 4 °C. Do not autoclave the infiltration buffer containing acetosyringone.

19. The confocal laser scanning microscope (CLSM) must have the capability to excite at the appropriate wavelengths. For example for the pEarleyGate 101 vector 514 nm for yellow fluorescent protein (YFP) is necessary.

20. After 7 days, a 120 mL culture should produce around 8–10 g of cells, fresh weight. The amount of biomass should be checked regularly and the volume of cells (usually 10–20 mL) used to subculture into a new flask adjusted accordingly.

21. Enzymes should be added to the digestion buffer just prior to use. Enzymes are easily solubilized by vigorous shaking in a 50 mL aliquot of digestion buffer, then adding to the solution containing the cells.

22. Maintain a cell to digestion buffer ratio of around 1:5 (w/v) for the digestion of cells. The amount of enzyme may need to be optimized if different plant material/cell cultures are employed. This combination and ratio of enzymes seems to be optimal for primary cell walls.

23. Rotate at the cells in the digestion buffer at the lowest possible speed; however, the cells must remain suspended.

24. The pellet comprising intact protoplasts is delicate and thus the supernatant should be carefully removed to avoid breaking protoplasts.

25. For the first few attempts at protoplast disruption, check the results using a light microscope with a 40× objective. This should reveal rupturing of at least 75 % of the protoplasts. The key to homogenization is ensuring enough mechanical stress to disrupt the protoplasts is used without destroying subcellular integrity. The number of strokes of the homogenizer should be calibrated using light microscopy.

26. This step should result in a yellow band sitting on the cushion (gradient buffer 1) of about 1–2 mm thick.

27. The removal of the supernatant to the top of the yellow membrane band will be a compromise between the quality of the resultant gradient and the disturbance of the cushion. Attempts to remove too much supernatant will likely result in a reduction in yield.

28. For the successful enrichment of Golgi membranes from other contaminating membranes on the cushion, such as mitochondria, it is not recommended to use a sucrose concentration less than about 1.0 M in gradient buffer 2.

29. The membrane band (discarded) found between gradient buffer 4 (0.1 M) and gradient buffer 3 (0.2 M) should be contain less material and be visibly thinner than the band representing the enriched Golgi membranes directly above the 1.0 M gradient buffer.

30. The incubation with KCl is used to remove electrostatic interactions of nonspecific peripheral proteins from the enriched Golgi membrane preparations.

31. The recommended ratio of trypsin to protein (w/w) is generally 1:20 or 50; however, this will depend on the type of trypsin being utilized. If using modified trypsin (not subject to autolysis), then recommended ratios can be used; however, if unmodified trypsin is used, we recommend using it at higher ratios. The advantage of unmodified trypsin is that autolytic products (peptides) can act as internal controls for sample handling and mass spectrometry.

32. Overnight digestion is likely excessive; however, the timing usually suits the standard sample processing workflow.

33. These volumes and concentrations are dependent on the liquid chromatography method being employed.

34. The analysis of a 1 μg complex peptide lysate by nanoflow LC-MS/MS should result in the high confidence identification of between 1000 and 2000 proteins.

35. To ensure only high confidence proteins and peptides are identified from fragmentation spectra, employ recommended protein and peptide cutoffs to ensure spectral match probabilities are <0.05 or false discovery rates are <1%. If using Mascot, an ions score value is provided for peptide matches and needs to be entered manually in the "Ions score or expect cut-off" box. Using such an approach, only high confidence peptide matches are used to define the resultant identified proteins. This means that even proteins identified by a single peptide can be regarded as high confidence matches, although a replicate experiment will be required to independently identify and confirm its existence in the sample.

36. There are further settings or views in Scaffold that can result in a final set of identified proteins that would not be considered high confidence even with the outlined settings (Fig. 2b). This includes the following settings in the "View" menu: "Show Entire Protein Clusters" and "Show Lower Scoring Matches". If requiring high confidence identifications, both should be unchecked.

37. Space the pots within the tray to ensure plants do not crowd each other as they mature.

38. Plants around 3–4 weeks after germination with a rosette diameter of approximately 80–100 mm seem to be the most suitable for infiltration.

39. Ensure the appropriate plant tissue in which the gene is most highly expressed is used. It is also important to use a proof-reading polymerase for the PCR reaction to avoid the introduction of sequence errors. For the same reason the number of PCR cycles should be kept to a minimum.

40. Competent Agrobacterium can be prepared in large batches and store 50 μL aliquots at –80 °C until required. About 1 μL (25–100 ng) of the plasmid preparation in water or TE buffer works for most *Agrobacterium* transformations. Ensure that plasmid preparations with higher DNA concentrations are diluted.

41. Make sure that your settings are adjusted for the gap size of the cuvette being employed and that the setting is appropriate for *Agrobacterium*.

42. It is essential to promptly add the LB-media to the mixture to avoid a decrease in transformation efficiency.

43. The volume largely depends on the efficiency rate of the competent *Agrobacterium*. Usually 50 μL is sufficient to produce positive colonies.

44. Ensure correct antibiotics are used for selection. Generally antibiotics for both the *Agrobacterium* host and plasmid are necessary. The *Agrobacterium* GV3101::pMP90 is resistant to gentamycin and rifampicin and the pEarleyGate vector series confer kanamycin resistance.

45. Some protocols do not include a wash step; however, we have found that this improves infiltration and protein expression. This could be due to the effect of antibiotics in the culture media.

46. The OD required depends on the expression of the protein of interest. Generally we have found that a lower OD (0.01–0.15) gives a better result and minimizes protein aggregates. However, low expressed proteins may require higher ODs. When co-infiltrating multiple constructs, for example, a gene of interest and an organelle marker protein, mix a combination of the two *Agrobacterium* solutions to achieve the final OD.

47. Start with the top leaves as they have fewer vascular bundles compared to the older leaves. We have found that slightly piercing the leaf by holding the syringe at a 45 ° angle while holding the leaf upright can assist infiltration.

48. It is important to use as little pressure as possible while injecting the *Agrobacterium* mixture so as to avoid punching holes and wounding the leaf.

49. Since not all proteins express with the same efficiency, expression should be monitored at various time points. Generally we monitor expression from 48 to 72 h.

50. Make sure the infiltrated area is chosen and avoid cutting too close to the region where the syringe was applied as this region will be wounded and will result in a high autofluorescence signal.

51. Most modern confocal microscopes are capable of performing simultaneous scans (fast, crosstalk between signals) and sequential scans (slow, but less crosstalk between signals). The use of simultaneous scanning enables the visualization of two signals without the need for fixation. This can be useful when examining the Golgi apparatus as they tend to move around the cell. While an emission range can be narrowed to reduce crosstalk when using simultaneous scanning, it will reduce the signal intensity. Sequential scanning is often essential to ensure the signal from both fluorophores is valid, but often fixation is required to ensure overlap. Such an approach is important if using fluorophores with similar emission wavelengths such as YFP and green fluorescent protein (GFP).

Acknowledgments

This work was funded by grants from the Australia Research Council (ARC) to the ARC Centre of Excellence in Plant Cell Walls [CE110001007] and the US Department of Energy, Office of Science, Office of Biological and Environmental Research, through contract DE-AC02-05CH11231 between Lawrence Berkeley National Laboratory and the US Department of Energy. J.L.H. is supported by an ARC Future Fellowship [FT130101165]. S.F.H. was supported by a research grant [VKR023371] from VILLUM FONDEN. We also wish to thank the UC Davis Proteomics Core Facility for sample analysis.

References

1. Morré DJ, Mollenhauer HH (2009) The Golgi apparatus: the first 100 years. Springer, New York

2. Scheller HV, Ulvskov P (2010) Hemicelluloses. Annu Rev Plant Biol 61:263–289

3. Harholt J, Suttangkakul A, Scheller HV (2010) Biosynthesis of pectin. Plant Physiol 153:384–395

4. Song W, Henquet MGL, Mentink RA et al (2011) N-glycoproteomics in plants: perspectives and challenges. J Proteomics 74:1463–1474

5. Rennie EA, Ebert B, Miles GP et al (2014) Identification of a sphingolipid alpha-glucuronosyltransferase that is essential for pollen function in Arabidopsis. Plant Cell 26:3314–3325

6. Wightman R, Turner S (2010) Trafficking of the plant cellulose synthase complex. Plant Physiol 153:427–432

7. Ordenes VR, Moreno I, Maturana D et al (2012) In vivo analysis of the calcium signature in the plant Golgi apparatus reveals unique dynamics. Cell Calcium 52:397–404

8. McFarlane HE, Watanabe Y, Yang WL et al (2014) Golgi- and trans-Golgi network-mediated vesicle trafficking is required for wax secretion from epidermal cells. Plant Physiol 164:1250–1260

9. Morré DJ, Mollenhauer HH (1974) The endomembrane concept: a functional integration of endoplasmic reticulum and Golgi apparatus. In: Robards AW (ed) Dynamic aspects of

plant infrastructure. McGraw-Hill, New York, pp 84–137

10. Morré DJ, Mollenhauer HH (1964) Isolation of Golgi apparatus from plant cells. J Cell Biol 23:295–305

11. Boevink P, Oparka K, Cruz SS et al (1998) Stacks on tracks: the plant Golgi apparatus traffics on an actin/ER network. Plant J 15:441–447

12. Akkerman M, Overdijk EJR, Schel JHN et al (2011) Golgi body motility in the plant cell cortex correlates with actin cytoskeleton organization. Plant Cell Physiol 52:1844–1855

13. Dunkley TPJ, Watson R, Griffin JL et al (2004) Localization of organelle proteins by isotope tagging (LOPIT). Mol Cell Proteomics 3:1128–1134

14. Nikolovski N, Rubtsov D, Segura MP et al (2012) Putative glycosyltransferases and other plant Golgi apparatus proteins are revealed by LOPIT proteomics. Plant Physiol 160:1037–1051

15. Parsons HT, Weinberg CS, Macdonald LJ et al (2013) Golgi enrichment and proteomic analysis of developing *Pinus radiata* xylem by free-flow electrophoresis. PLoS One 8:e84669

16. Parsons HT, Christiansen K, Knierim B et al (2012) Isolation and proteomic characterization of the Arabidopsis Golgi defines functional and novel targets involved in plant cell wall biosynthesis. Plant Physiol 159:12–26

17. Parsons HT, González Fernández-Niño SM, Heazlewood JL (2014) Separation of the plant Golgi apparatus and endoplasmic reticulum by free-flow electrophoresis. In: Jorrín Novo JV, Komatsu S, Weckwerth W, Weinkoop S (eds) Plant proteomics: methods and protocols, vol 1072, 2nd edn. Humana Press, New York, pp 527–539

18. Forsmark A, Rossi G, Wadskog I et al (2011) Quantitative proteomics of yeast post-Golgi vesicles reveals a discriminating role for Sro7p in protein secretion. Traffic 12:740–753

19. Gilchrist A, Au CE, Hiding J et al (2006) Quantitative proteomics analysis of the secretory pathway. Cell 127:1265–1281

20. Mast S, Peng L, Jordan TW et al (2010) Proteomic analysis of membrane preparations from developing *Pinus radiata* compression wood. Tree Physiol 30:1456–1468

21. Zeng W, Jiang N, Nadella R et al (2010) A glucurono(arabino)xylan synthase complex from wheat contains members of the GT43, GT47, and GT75 families and functions cooperatively. Plant Physiol 154:78–97

22. Nikolovski N, Shliaha PV, Gatto L et al (2014) Label-free protein quantification for plant Golgi protein localization and abundance. Plant Physiol 166:1033–1043

23. Murashige T, Skoog F (1962) A revised medium for rapid growth and bio assays with tobacco tissue cultures. Physiol Plant 15:473–497

24. Earley KW, Haag JR, Pontes O et al (2006) Gateway-compatible vectors for plant functional genomics and proteomics. Plant J 45:616–629

25. Gene Ontology Consortium (2004) The Gene Ontology (GO) database and informatics resource. Nucleic Acids Res 32:D258–D261

26. Heazlewood JL, Verboom RE, Tonti-Filippini J et al (2007) SUBA: the Arabidopsis subcellular database. Nucleic Acids Res 35:D213–D218

27. Schneider CA, Rasband WS, Eliceiri KW (2012) NIH Image to ImageJ: 25 years of image analysis. Nat Methods 9:671–675

28. May MJ, Leaver CJ (1993) Oxidative stimulation of glutathione synthesis in *Arabidopsis thaliana* suspension cultures. Plant Physiol 103:621–627

29. Tanz SK, Castleden I, Hooper CM et al (2013) SUBA3: a database for integrating experimentation and prediction to define the SUBcellular location of proteins in Arabidopsis. Nucleic Acids Res 41:D1185–D1191

30. Zybailov B, Mosley AL, Sardiu ME et al (2006) Statistical analysis of membrane proteome expression changes in *Saccharomyces cerevisiae*. J Proteome Res 5:2339–2347

31. Ebert B, Rautengarten C, Guo X et al (2015) Identification and characterization of a Golgi-localized UDP-xylose transporter family from Arabidopsis. Plant Cell 27:1218–1227

Chapter 9

High-Content Analysis of the Golgi Complex by Correlative Screening Microscopy

Manuel Gunkel, Holger Erfle, and Vytaute Starkuviene

Abstract

The Golgi complex plays a central role in a number of diverse cellular processes, and numerous regulators that control these functions and/or morphology of the Golgi complex are known by now. Many of them were identified by large-scale experiments, such as RNAi-based screening. However, high-throughput experiments frequently provide only initial information that a particular protein might play a role in regulating structure and function of the Golgi complex. Multiple follow-up experiments are necessary to functionally characterize the selected hits. In order to speed up the discovery, we have established a system for correlative screening microscopy that combines rapid data collection and high-resolution imaging in one experiment. We describe here a combination of wide-field microscopy and dual-color direct stochastical optical reconstruction microscopy (*d*STORM). We apply the technique to simultaneously capture and differentiate alterations of the *cis-* and *trans-*Golgi network when depleting several proteins in a singular and combinatorial manner.

Key words Golgi complex, *d*STORM, Correlative microscopy, Golgin-97, AP3S1

1 Introduction

In recent years, microscope systems for high-throughput (HTS) and high-content (HCS) screening have emerged. These systems are capable of automatically preparing, acquiring, and processing huge sample numbers in a quick and robust manner. Fluorescence screening microscopy has been successfully applied to analyze virtually every cellular process; and more than several thousand large-scale screens have been published to date. Owning to the multiple functions of the Golgi complex—regulation of the secretory membrane trafficking, signal transduction, protein and lipid glycosylation, unfolded protein response, and many others [1]—more than 30 screens have been performed to address one or another Golgi complex-related function in a systematic manner. A comprehensive list of membrane trafficking regulators acting in anterograde and retrograde transport routes to, through, and from the Golgi

William J. Brown (ed.), *The Golgi Complex: Methods and Protocols*, Methods in Molecular Biology, vol. 1496,
DOI 10.1007/978-1-4939-6463-5_9, © Springer Science+Business Media New York 2016

complex is compiled by now [2–7]. Protein glycosylation [8] and protease activities [9] in the Golgi complex were also studied systematically. The organization of the Golgi complex changes upon diverse conditions: cell division, viral infection, perturbed flow of material, or altered cytoskeleton organization. Furthermore, changed Golgi complex was found in a number of pathological conditions, such as Alzheimer [10]. In addition, screens dedicated to analyze Golgi morphology are gaining in popularity [11]. On the other hand, such type of screens are fairly easy to establish due to advances in labeling, visualization and quantification of the altered Golgi complex by fluorescence screening microscopy and automated image analysis [12, 13]. Wide-field or confocal microscope-based screens are usually performed at a low resolution with 10× or 20× objectives [7, 11] in order to collect a high number of cells and identify phenotypic changes in cell populations on a statistically sound basis. Usually, that is sufficient to detect the regulatory molecules that induce profound alterations marked as "fragmentation," "disappearance," or "condensation" of the Golgi [4, 11]. The alterations that take place on the level of individual cisternae, vesicle budding or cargo sorting stay largely uncharacterized that way. Morphological and functional details of the Golgi complex can be obtained by various high-resolution microscopy methods: STED (Stimulated Emission Depletion) [14], SCLIM (super-resolution confocal live imaging microscopy) [15], 4Pi [16], and *d*STORM [17]. For simultaneous acquisition of spatio-temporal resolved information, such as imaging of fast trafficking events with high morphological resolution, several microscopy methods to image one and the same specimen need to be combined [18]. For example, carrier budding at the trans-Golgi network (TGN) was visualized and followed by correlation of light and electron microscopy (CLEM) [19, 20].

All in all, correlative microscopy provides a unique possibility to combine the particular strength of each microscopic approach and compensate for its specific limitations. In order to achieve a balance between high throughput, time efficiency, and data depth, we have recently developed a correlative screening microscopy platform [17]. The sample is initially imaged by low-resolution screening microscopy, objects of the potential interest are examined manually or automatically and their positions relocated to high-resolution microscope system of choice (Fig. 1). This can be achieved either by referencing the sample based on specific markers on the slide itself or by using the fluorescence signal of the specimen. In our case the individual cells and their spatial distribution serve as the reference points, which can be identified in any microscope system and thus be used as anchor points for a coordinate transformation.

Here, we demonstrate advantages of correlative wide-field and *d*STORM microscopy to assess alterations of *cis*- and *trans*-Golgi when several Golgi-localized proteins are downregulated by RNAi (RNA interference). One of them is Golgi matrix protein, golgin-97

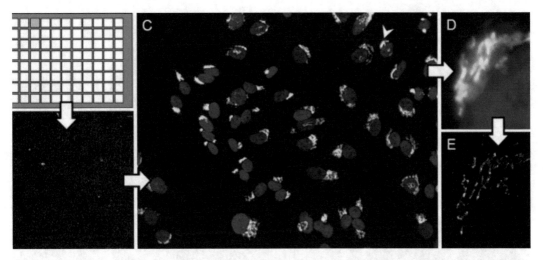

Fig. 1 Schematic overview of the workflow of the correlative microscopy. A multi-well plate (**a**) is imaged in multiple positions/well (**b**). Scale bar in (**b**): 500 μm. Within these single images, representative cells are selected for super-resolution imaging (*white arrowhead* in **c**). Scale bar in (**c**): 50 μm. The sample is transferred to a super-resolution imaging capable microscope and referenced in order to find back the marked positions. For each marked cell, a regular wide-field image (63× magnification) of the Golgi complex is acquired (**d**) (scale bar: 5 μm), after that, the same region is acquired and reconstructed in super-resolution mode as shown in (**e**)

[21], which mediates bidirectional endosomes-to-TGN trafficking [22, 23]. It was reported, that depletion of golgin-97 by the microinjected antibody induced strong fragmentation of the whole Golgi complex [23]. siRNA-mediated downregulation removed more than 75 % of the protein, but little changes on the Golgi complex were observed by a wide-field microscopy [22, 23]. The observation was attributed to the fact that the remaining amounts of golgin-97 are sufficient to maintain the structure of the Golgi complex. In contrast, *d*STORM imaging of cells after downregulation of golgin 97 with the respective siRNAs revealed a disorganization of the compact Golgi complex (Fig. 2), indicating that the morphology of the organelle and the expression levels of golgin-97 are closely related. Another analyzed protein, AP3S1, is a subunit of AP-3 adaptor complex, which localizes to the TGN and endosomes [24, 25]. AP-3 plays a role in TGN–lysosome and endosome–lysosome trafficking [25–27]. Whether RNAi-mediated depletion of AP-3 induces changes of the Golgi complex was not tested before. The alterations are plausible as downregulation of the other TGN-localized adaptor complex, namely, AP-1, induces a strong fragmentation of the Golgi complex [28]. Similar to golgin-97, depletion of AP3S1 induced mild changes of the Golgi that are hardly visible in a wide-field modus, but are clearly identifiable in *d*STORM. Simultaneous downregulation of golgin-97 and AP3S1 induced an additive effect: extension and simultaneous fragmentation of the Golgi ribbon. This finding suggests a possibility that both proteins

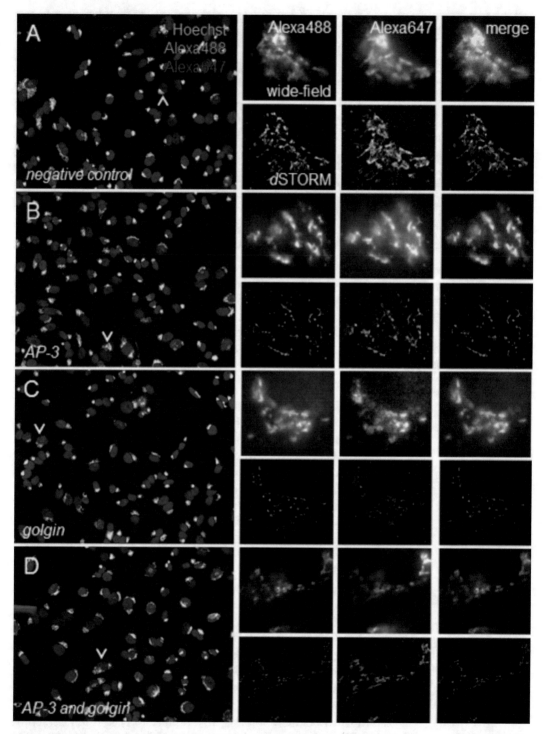

Fig. 2 *d*STORM reveals mild alterations of the Golgi complex induced by the downregulation of golgin-97 and AP3S1. Hela cells were transfected with siRNAs targeting golgin-97 and AP3S1, incubated for 48 h, fixed and stained with the Golgi markers labeling *cis*- and *trans*-network (Subheading 2.1, **item 2**). Images taken by a wide-field microscopy are shown on the *left side* of the figure (scale bar: 50 μm). The selected cells (marked by the *arrowheads*) were repeatedly imaged by a wide-field microscopy and *d*STORM (*right side* of the figure, scale bars: 5 μm) for resolving of the potential changes of the Golgi complex under the conditions of RNAi. *Arrows* indicate differential distribution of *cis*- and *trans*-Golgi markers

might be functionally related more that it is currently anticipated. The alterations are recognizable in a wide-field microscopy, however, differential redistribution of *cis-* and *trans-*Golgi markers are visible only in higher resolution (indicated by arrows in Fig. 2). In cells treated with the negative siRNA Golgi complex remains compact with well-resolved *cis-* and *trans-*sides by *d*STORM.

2 Materials

2.1 Assay for the Golgi Complex Integrity

1. HeLa cells ATCC (CCL-2) growing in DMEM containing 10% fetal calf serum, 2 mM glutamine, 100 U/ml penicillin, and 100 µg/ml streptomycin in 8-well µ-slide (Ibidi, Germany).

2. Mouse monoclonal anti-GM130 (BD Transduction Laboratories).

3. Sheep polyclonal anti-TNG46 antibodies (AbD Serotec).

4. Anti-mouse and anti-sheep antibodies conjugates to Alexa Fluor® 532 and Alexa Fluor® 647 (Thermo Fisher Scientific) (*see* **Note 1**).

5. Hoechst 33342 at the final concentration 0.1 µg/ml (in PBS).

6. 0.1% Triton X-100 in PBS.

7. 3% paraformaldehyde in PBS, pH 7.4.

2.2 dSTORM Imaging

1. Switching buffer consists of 100 mM mercaptoethanol solution in PBS.

3 Methods

3.1 RNAi to Observe Changes of the Golgi Complex

1. Plate 7000–9000 HeLa cells/well in the 8-well µ-slide containing 300 µl of the growth medium/well and incubate for 24 h at 37 °C.

2. Transfect cells with 50 nM end concentration of single or combined siRNAs using Lipofectamine 2000 or equivalent reagent according the protocol provided by the manufactures (*see* **Note 2**). Incubate transfected cells for 48 h before fixation with 3% paraformaldehyde in PBS for 20 min at RT and permeabilize with Triton X-100 for 5 min at RT prior immunofluorescence staining (*see* **Note 3**).

3.2 Immunofluore scence Methods

1. Endogenous proteins localized to the *cis-* and *trans-*Golgi are visualized by immunostaining with mouse monoclonal anti-GM130 and sheep polyclonal anti-TNG46 antibodies, respectively, and anti-mouse antibodies conjugates to Alexa Fluor® 532 and anti-sheep antibodies conjugated to Alexa Fluor® 647.

2. Stain cell nuclei with Hoechst 33342 for 5 min at RT.

3.3 Wide-Field Microscopy

1. For the fast imaging of the sample, an Olympus IX81 ScanR system (Olympus, Germany) was used with a 20× magnifying objective (Olympus UPlanSApo, NA 0.75).

2. The gradient based software autofocus was done in the Hoechst color channel (405 nm for excitation) in two steps to find the plane of the nuclei: first, a coarse search for a maximum of the gradient function was performed over an axial range of ±60 μm in 6 μm steps. Then at the position of the maximum fine focusing was performed the same way over an axial range of ±10 μm in 0.8 μm steps.

3. Three color channels for Hoechst, Alexa Fluor® 532 and Alexa Fluor® 647 staining were recorded at center wavelengths for detection of 465, 532 nm and 647 nm to visualize the nucleus, *cis-* and *trans-*Golgi.

4. In every well, cells were imaged with an overlap between adjacent images of 10 % resulting in 396 sub-positions in order to cover the central area of each well (*see* **Note 4**).

3.4 Cell Referencing and Coordinate Transfer

1. The cells with the altered Golgi were marked manually based upon visual inspection of wide-field images in ImageJ (*see* **Note 5**). The coordinates of the cells along with the according source image names were saved in a text file.

2. The sample was transferred to a Leica SP5 confocal microscope with an additional localization microscopy unit (*see* **Note 6**).

3. Cell coordinates were matched by using nuclei as a referencing structure. In the center of each well, one wide-field image was acquired of the Hoechst stain. These images (target images) were scaled and rotated in order to match the previously acquired images from the Olympus IX81 microscope (source images). Each image was normalized and cross-correlated via an ImageJ macro with all the previously acquired images of each well. At the position of the best match between source and target image the resulting cross-correlated image shows a global maximum. This position is recorded along with the respective source and target image and describes the offset between these two images (*see* **Note 7**).

4. From the list of associated source and target images along with the offset between these obtained by the cross-correlation, the three pairs providing the best matches (highest maxima) were used as reference points for a coordinate transfer function. By this function the coordinates of the identified cells of interest determined in the images of the IX81 microscope system (stage position of the image + position of the cell within the image) could be transferred to the SP5 system (position of the motorized stage in order to put the cell of interest in the center of the field of view).

3.5 dSTORM Imaging

1. For *d*STORM imaging of the Alexa Fluor® 647 and Alexa Fluor® 532 a modified Leica SP5 system with custom adapted wide-field detection and laser illumination was used (Fig. 3) (*see* **Note 8**).

2. For *d*STORM sample illumination a custom build TIRF laser illumination was used (*see* **Note 9**). This comprises a laser combiner box (wavelengths 405 nm, 488 nm, 515 nm, 561 nm, and 647 nm), beam expansion and a movable mirror integrated in a microbench coupled to the backward illumination port (P2 in Fig. 3) of the microscope stand. Images were acquired by a Hamamatsu OrcaFlash4.0 sCMOS camera mounted to the sideward wide-field port (P3 in Fig. 3) of the microscope stand.

3. PBS medium in the fixed sample was replaced with switching buffer shortly before imaging (Subheading 2.2).

4. Based on the coordinate transfer function (Subheading 3.4, **step 4**) a coordinate list for the cells of interest was generated which could be loaded in the LAS-AF software controlling the Leica SP5. Within this software the respective cell positions could be selected. This triggered the microscope stage to move to the appropriate position.

5. Prior to *d*STORM imaging (and thus bleaching of the staining) wide-field images of both channels were acquired at low illumination intensities at the same lateral and axial position in for comparison.

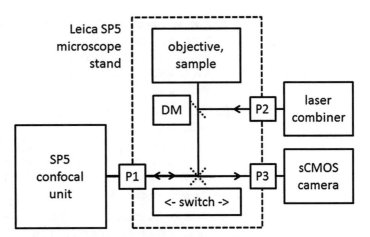

Fig. 3 Schematic of the setup used for super-resolution microscopy. A Leica SP5 confocal microscope is used for super-resolution microscopy, which comprises an additional dichroic filter changer (DM) and a switch for change between confocal excitation and detection or wide-field detection. For the confocal operation, the dichroic filter is moved out of the beam path and the switch set to the confocal path (port P1). For super-resolution microscopy, the dichroic is placed in the beam path. Additional laser illumination is coupled into the microscope stand via port P2. The switch is set to wide-field detection via port P3 and the fluorescence signal is registered by a sCMOS camera

6. For *d*STORM imaging of the Alexa Fluor® 647 stain, the 647 nm laser intensity was raised to 140 mW and an imaging sequence of 3000 images with 50 ms integration time for each frame was acquired.

7. Laser and filter settings were switched to 515 nm excitation (150 mW) and an image sequence for the Alexa Fluor® 532 stain was acquired using the same imaging parameters.

8. Images were reconstructed with the open source ImageJ plugin QuickPALM [29]. This plugin analyzes the recorded image sequence frame by frame and determines the center positions of the individual single molecule fluorescence signals within each frame. The analysis was started by selecting Analyze > QuickPALM > Analyze Particles in ImageJ. Various parameters were set, like minimum signal-to-noise ratio (SNR): 4, width of the estimated single molecule signal distribution: 3 pixel and pixel size of the camera: 106 nm, in agreement to the used fluorophores and camera, respectively. QuickPALM offers further possibilities for 3D estimation and image rendering, but these functions have not been used in the presented application, and therefore are not described. Position data is stored in a file and can be represented as 2D reconstructed image like shown in Fig. 1e (*see* **Note 10**).

4   Notes

1. The secondary fluorophores were chosen to be compatible with and well-resolved by both wide-field microscopy and *d*STORM. The indicated dyes are not the only possible pair as *d*STORM is applicable to a broad range of fluorescent dyes and proteins [17, 30–32].

2. Transfection method largely depends on the used cells and the molecules that needs to be transfected (e.g., siRNAs or cDNAs), and is not limited to the usage of Lipofectamine 2000. For example, similar or even higher transfection rates for siRNAs in HeLa cells can be achieved with Lipofectamine RNAiMAX (Thermo Fisher Scientific) or HiPerFect (Qiagen).

3. Permeabilization of cells can be achieved via different methods and the choice of the method depends on the intracellular epitopes. For example, anti-TGN46 antibody can be used in cells permeabilized with Triton and methanol, whereas, anti-GM130 (both antibodies from the provider described here) does not result in a good staining in cells fixed with methanol.

4. Wide-field imaging of all samples needs to be done at the same well positions to secure consistent object registration and coordinate transfer.

5. Manual or automatic identification of cells of interest can be performed using any software as long as the position information within the source images can be retrieved.

6. Any type of wide-field or confocal microscope can be used as long as it has a motorized sample stage and is capable to perform automatic image acquisition.

7. For referencing the images of both microscope systems have to be matched in terms of scaling, offset, and rotation. In some cases anisotrope or other distortions have also to be accounted for, but usually an affine transformation is sufficient in order to describe the relationship between both images. Generally only sufficiently large parts of the images have to overlap in order to determine the transfer function.

8. For single molecule localization based super-resolution microscopy the usage of Leica SP5 system is not obligatory. Usually, a wide-field laser illumination with sufficient power (e.g., 647 nm laser of 150 mW) to induce fluorophore photoswitching can be used. In addition, a wide-field detection with improved signal-to-noise ratio is needed (e.g., EM-CCD camera).

9. In order to remove background signal coming from off-focus planes, restricting the plane of illumination by SPIM [33] or TIRF is recommended.

10. Image sequences can also be analyzed by other software or ImageJ plugins, such as ThunderSTORM [34], which fits a model function to the intensity distribution. As a result, QuickPALM yields a quick result for the inspection and the quality control of the experiment. ThunderSTORM provides more precise localization of the single fluorophores. In ThunderSTORM, camera parameters have to be set prior to analysis. For the analysis itself, default parameters can be taken. For in depth description please refer to the documentation. A list of coordinates is also generated which can be post-processed in order to get rid of coordinates with insufficient localization accuracy or are too broad to be a single molecule signal, for instance.

Acknowledgements

This work was supported by the grant "Methoden für die Lebenswissenschaften" of the Stiftung Baden-Württemberg (Nr. P-LS-SPII/11) and CellNetworks-Cluster of Excellence, Heidelberg University (EXC81). We thank Jamila Begum for the preparation of the samples.

References

1. Wilson C, Venditti R, Rega LR, Colanzi A, D'Angelo G, De Matteis MA (2011) The Golgi apparatus: an organelle with multiple complex functions. Biochem J 433:1–9

2. Galea G, Bexiga MG, Panarella A, O'Neill ED, Simpson JC (2015) A high-content screening microscopy approach to dissect the role of Rab proteins in Golgi-to-ER retrograde trafficking. J Cell Sci 128:2339–2349

3. Anitei M, Chenna R, Czupalla C, Esner M, Christ S, Lenhard S, Korn K, Meyenhofer F, Bickle M, Zerial M, Hoflack B (2014) A high-throughput siRNA screen identifies genes that regulate mannose 6-phosphate receptor trafficking. J Cell Sci 127:5079–5092

4. Simpson JC, Joggerst B, Laketa V, Verissimo F, Cetin C, Erfle H, Bexiga MG, Singan VR, Heriche JK, Neumann B, Mateos A, Blake J, Bechtel S, Benes V, Wiemann S, Ellenberg J, Pepperkok R (2012) Genome-wide RNAi screening identifies human proteins with a regulatory function in the early secretory pathway. Nat Cell Biol 14:764–774

5. Kondylis V, Tang Y, Fuchs F, Boutros M, Rabouille C (2011) Identification of ER proteins involved in the functional organisation of the early secretory pathway in drosophila cells by a targeted RNAi screen. PLoS One 6, e17173

6. Lisauskas T, Matula P, Claas C, Reusing S, Wiemann S, Erfle H, Lehmann L, Fischer P, Eils R, Rohr K, Storrie B, Starkuviene V (2012) Live-cell assays to identify regulators of ER-to-Golgi trafficking. Traffic 13:416–432

7. Starkuviene V, Liebel U, Simpson JC, Erfle H, Poustka A, Wiemann S, Pepperkok R (2004) High-content screening microscopy identifies novel proteins with a putative role in secretory membrane traffic. Genome Res 14:1948–1956

8. Goh GY, Bard FA (2015) RNAi screens for genes involved in Golgi glycosylation. Methods Mol Biol 1270:411–426

9. Coppola JM, Hamilton CA, Bhojani MS, Larsen MJ, Ross BD, Rehemtulla A (2007) Identification of inhibitors using a cell-based assay for monitoring Golgi-resident protease activity. Anal Biochem 364:19–29

10. Joshi G, Wang Y (2015) Golgi defects enhance APP amyloidogenic processing in Alzheimer's disease. Bioessays 37:240–247

11. Chia J, Goh G, Racine V, Ng S, Kumar P, Bard F (2012) RNAi screening reveals a large signaling network controlling the Golgi apparatus in human cells. Mol Syst Biol 8:629

12. Miller VJ, McKinnon CM, Mellor H, Stephens DJ (2013) RNA interference approaches to examine Golgi function in animal cell culture. Methods Cell Biol 118:15–34

13. Starkuviene V, Seitz A, Erfle H, Pepperkok R (2008) Measuring secretory membrane traffic: a quantitative fluorescence microscopy approach. Methods Mol Biol 457:193–201

14. Erdmann RS, Takakura H, Thompson AD, Rivera-Molina F, Allgeyer ES, Bewersdorf J, Toomre D, Schepartz A (2014) Super-resolution imaging of the Golgi in live cells with a bioorthogonal ceramide probe. Angew Chem Int Ed Engl 53:10242–10246

15. Kurokawa K, Ishii M, Suda Y, Ichihara A, Nakano A (2013) Live cell visualization of Golgi membrane dynamics by super-resolution confocal live imaging microscopy. Methods Cell Biol 118:235–242

16. Perinetti G, Muller T, Spaar A, Polishchuk R, Luini A, Egner A (2009) Correlation of 4Pi and electron microscopy to study transport through single Golgi stacks in living cells with super resolution. Traffic 10:379–391

17. Flottmann B, Gunkel M, Lisauskas T, Heilemann M, Starkuviene V, Reymann J, Erfle H (2013) Correlative light microscopy for high-content screening. Biotechniques 55:243–252

18. Hubner B, Cremer T, Neumann J (2013) Correlative microscopy of individual cells: sequential application of microscopic systems with increasing resolution to study the nuclear landscape. Methods Mol Biol 1042:299–336

19. Polishchuk EV, Di Pentima A, Luini A, Polishchuk RS (2003) Mechanism of constitutive export from the Golgi: bulk flow via the formation, protrusion, and en bloc cleavage of large trans-Golgi network tubular domains. Mol Biol Cell 14:4470–4485

20. Polishchuk EV, Polishchuk RS (2013) Analysis of Golgi complex function using correlative light-electron microscopy. Method Cell Biol 118:243–258

21. Griffith KJ, Chan EK, Lung CC, Hamel JC, Guo X, Miyachi K, Fritzler MJ (1997) Molecular cloning of a novel 97-kd Golgi complex autoantigen associated with Sjogren's syndrome. Arthritis Rheum 40:1693–1702

22. Lock JG, Hammond LA, Houghton F, Gleeson PA, Stow JL (2005) E-cadherin transport from the trans-Golgi network in tubulovesicular carriers is selectively regulated by Golgin-97. Traffic 6:1142–1156

23. Lu L, Tai G, Hong W (2004) Autoantigen Golgin-97, an effector of Arl1 GTPase, participates in traffic from the endosome to the trans-Golgi network. Mol Biol Cell 15:4426–4443

24. Dell'Angelica EC, Ohno H, Ooi CE, Rabinovich E, Roche KW, Bonifacino JS (1997) AP-3: an adaptor-like protein complex with ubiquitous expression. EMBO J 16:917–928

25. Peden AA, Oorschot V, Hesser BA, Austin CD, Scheller RH, Klumperman J (2004) Localization of the AP-3 adaptor complex defines a novel endosomal exit site for lysosomal membrane proteins. J Cell Biol 164:1065–1076

26. Rous BA, Reaves BJ, Ihrke G, Briggs JA, Gray SR, Stephens DJ, Banting G, Luzio JP (2002) Role of adaptor complex AP-3 in targeting wild-type and mutated CD63 to lysosomes. Mol Biol Cell 13:1071–1082

27. Le Borgne R, Alconada A, Bauer U, Hoflack B (1998) The mammalian AP-3 adaptor-like complex mediates the intracellular transport of lysosomal membrane glycoproteins. J Biol Chem 273(45):29451–29461

28. Rollason R, Korolchuk V, Hamilton C, Schu P, Banting G (2007) Clathrin-mediated endocytosis of a lipid-raft-associated protein is mediated through a dual tyrosine motif. J Cell Sci 120:3850–3858

29. Henriques R, Lelek M, Fornasiero EF, Valtorta F, Zimmer C, Mhlanga MM (2010) QuickPALM: 3D real-time photoactivation nanoscopy image processing in ImageJ. Nat Methods 7:339–340

30. Gunkel M, Erdel F, Rippe K, Lemmer P, Kaufmann R, Hormann C, Amberger R, Cremer C (2009) Dual color localization microscopy of cellular nanostructures. Biotechnol J 4:927–938

31. van de Linde S, Heilemann M, Sauer M (2012) Live-cell super-resolution imaging with synthetic fluorophores. Annu Rev Phys Chem 63:519–540

32. Klein T, van de Linde S, Sauer M (2012) Live-cell super-resolution imaging goes multicolor. Chembiochem 13:1861–1863

33. Cella Zanacchi F, Lavagnino Z, Perrone Donnorso M, Del Bue A, Furia L, Faretta M, Diaspro A (2011) Live-cell 3D super-resolution imaging in thick biological samples. Nat Methods 8:1047–1049

34. Ovesny M, Krizek P, Borkovec J, Svindrych ZK, Hagen GM (2014) ThunderSTORM: a comprehensive ImageJ plug-in for PALM and STORM data analysis and super-resolution imaging. Bioinformatics 30:2389–2390

Chapter 10

Activity Detection of GalNAc Transferases by Protein-Based Fluorescence Sensors In Vivo

Lina Song, Collin Bachert, and Adam D. Linstedt

Abstract

Mucin-type O-glycosylation occurring in the Golgi apparatus is an important protein posttranslational modification initiated by up to 20 GalNAc-transferase isozymes with largely distinct substrate specificities. Regulation of this enzyme family affects a vast array of proteins transiting the secretory pathway and misregulation causes human diseases. Here we describe the use of protein-based fluorescence sensors that traffic in the secretory pathway to monitor GalNAc-transferase activity in living cells. The sensors can either be "pan" or isozyme specific.

Key words O-glycosylation, GalNAc transferase, Fluorescent biosensor, Fluorescence-activating protein

1 Introduction

Glycosylation enzymes change the surface characteristics of their protein substrates by appending or modifying carbohydrate chains termed glycans. Glycans impact protein solubility, stability, and interactions [1–3]. Mucin-type O-glycosylation is a large and important subgroup defined by the initial reaction in which N-acetylgalactosamine (GalNAc) is transferred to the hydroxyl group of serine or threonine (and possibly tyrosine) by UDP-N-acetyl-α-D-galactosamine polypeptide N-acetylgalactosaminyl transferases (GalNAc transferases). This occurs on secretory cargo as it passes through the Golgi apparatus and other Golgi-localized enzymes extend the glycan by further additions of individual monosaccharides. While one or two isoforms exist for each of the enzymes mediating chain extension, there are up to 20 distinct genes encoding GalNAc-transferase isozymes in humans, each with at least partial substrate selectivity [4]. Collectively, the isozymes modify a vast number of substrates and they are already linked to a significant number of medical syndromes [1, 5–8].

William J. Brown (ed.), *The Golgi Complex: Methods and Protocols*, Methods in Molecular Biology, vol. 1496,
DOI 10.1007/978-1-4939-6463-5_10, © Springer Science+Business Media New York 2016

Despite their biological and medical significance, assays have been lacking that monitor GalNAc-transferase activity in living cells and there are no known inhibitors. Therefore, we developed cell-based fluorescence sensors that are particularly sensitive to inhibition of GalNAc-transferase activity [9, 10]. They are transfectable constructs encoding proteins that traffic to the cell surface and become fluorescent if their glycosylation is inhibited. In brief, each sensor has a fluorescence activating protein domain (FAP) followed by a linker to a blocking domain that occludes dimerization necessary for binding of the dye malachite green to the FAP (Fig. 1). The dye itself is nonfluorescent and only becomes fluorescent when bound to the FAP [11]. Within the linker is a glycan acceptor site next to a furin protease site. Because GalNAc-transferases are localized to the cis/medial Golgi and furin is localized to the TGN, glycan addition occurs before, and therefore sterically blocks, processing by furin leaving the sensor inactive. Extending enzymes also act before furin and contribute to glycan-masking (*see* **Note 1**). However, if the GalNAc-transferase activity is inhibited, the entire glycan is absent and furin cleaves the linker releasing the blocking domain and allowing dye activation. The sensors are ratiometric because they also contain a fluorescent protein domain.

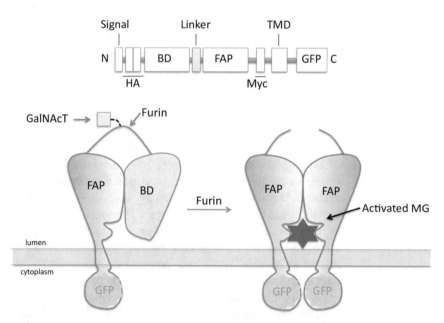

Fig. 1 Sensor schematic. The linear diagram (not to scale) depicts the N-terminal cleaved signal sequence, tandem HA epitopes, blocking domain (BD), linker, fluorescence-activating protein (FAP), myc epitope, transmembrane domain (TMD), and fluorescent protein domain (GFP). The cartoon shows an intact sensor on the *left* with adjacent GalNAc-transferase and furin sites in the linker such that GalNAc addition masks the furin site. Reduced GalNAc-T activity allows furin cleavage releasing the BD. As shown on the *right*, the FAP then dimerizes and binds and activates the dye malachite green (MG). The signal is read as the MG–GFP ratio

There are two critical aspects regarding sensor function. First, the sequence of the glycan acceptor site can either be "pan" or isozyme specific. The pan-specific version contains a minimal sequence recognized by GalNAc-transferases [12]. Although not confirmed for every isozyme, it should be recognized as a substrate by most if not all [13, 14]. For the isozyme-specific versions, the sequences are tailor-made for selective recognition by individual isozymes. At present only T2- and T3-specific versions are available. Their acceptor sequences were derived from known T2 or T3 substrates and then modified to improve selectivity [6, 10, 13, 15, 16]. The second critical aspect is that the starting or ground state signal can be significantly greater than true background (where true background is determined by mutating the furin site). This is advantageous if the desire is to assay enzyme activation. A less than optimal glycosylation site results in incomplete glycosylation and therefore an increased starting signal. Conditions upregulating enzyme activity will then lower the signal. Thus, although the sensors were designed to assay for inhibitors, they can be further modified and used to monitor changes in GalNAc-transferase activity in either direction.

2 Materials

2.1 Cell Culture

1. HEK293 cells.

2. HEK293 growth medium: minimal essential medium containing 10% fetal bovine serum and 100 IU/ml penicillin–streptomycin. Store at 4 °C. Use in 5% humidified CO_2 incubator at 37 °C.

3. Malachite Green-11p-NH2 (MG11P) (Sharp Edge Labs, Pittsburgh, PA) (*see* **Note 2**): Reconstitute at 110 μM (1000× stock) in 95% methanol, 5% acetic acid. Store at 4 °C in the dark.

4. MEM without phenol red: Adjust to 110 nM MG11P just before use.

5. EDTA/PBS: Reconstitute 8 g NaCl, 0.2 g KCl, 0.916 g Na_2HPO_4, and 0.2 g KH_2PO_4 in 1 L water at pH 7.3. Add 2.081 g EDTA (5 mM final concentration) and adjust pH, if necessary. Adjust to 110 nM MG11P just before use.

2.2 Confocal Assay

1. Live cell imaging chamber (or glass bottom dishes) and inverted confocal microscope equipped with a 40× oil immersion objective and 488 nm and 633 nm filter sets (*see* **Note 3**).

2. Image analysis software (e.g., ImageJ, http://imagej.nih.gov/ij/).

2.3 Flow Cytometry Assay

1. 96-well flat-bottom plastic dishes.

2. EDTA/PBS containing 110 nM MG11P.

3. Flow cytometer equipped to read at 488 nm and 640 nm (*see* **Note 3**) and accompanying software.

3 Methods

Below are two protocols for use of the GalNAc transferase biosensors. The first uses live-cell fluorescence microscopy and the second employs flow cytometry. Flow cytometry is preferred because a large number of cells are readily quantified for each condition and multi-well dishes can be used allowing a large number of conditions to be tested (e.g., high-throughput screening). The sensors can also be analyzed using immunoblotting if biochemical verification is desired (*see* **Note 4**). As a starting point, each protocol requires at least 3 cell lines: experimental, negative control and positive control. The experimental cell line(s) stably express the GalNAc-T biosensor(s) of interest. The linker sequence in these will contain either the pan-specific glycan acceptor site or one of the isozyme specific acceptor sites. The signal produced can be compared for differing conditions and normalized to the controls for comparison with other sensor signals. For the negative control, either untransfected cells or, optimally, cells stably expressing a matched version of the sensor with a mutated furin acceptor site (ΔFur) is used. This cell line will establish the true background, i.e., the signal in the absence of any cleavage, which theoretically corresponds to complete glycosylation of all sensor molecules. The positive control is a cell line expressing a matched sensor with a mutated glycan acceptor site (ΔGly). This cell line will establish the maximal possible signal given that it cannot be glycosylated. The protocols are described for use of the preexisting HEK293 cell lines expressing pan-, T2-, and T3-specific sensors and their matched controls. However, the linker sequence can be modified to produce new specificities (*see* **Note 5**). Also, the assay should work in most, if not all, cell lines but each cell type may require minor modifications in cell handling. After processing in the Golgi the sensor accumulates on the cell surface. It takes a minimum of 3 h to develop a strong surface signal, so, after a test condition, a period of at least this duration should be allowed before the cells are analyzed [10].

3.1 Assay GalNAc-T Activity Using Biosensor and Confocal Microscopy

1. Pass the cell lines (ΔFur, biosensor, ΔGly) to achieve 50 % confluence on coverslips for imaging chamber or in glass bottom dishes and incubate for 24 h (*see* **Note 6**).

2. Transfer cells to pre-warmed MEM without phenol red and containing 110 nM MG11P dye.

3. Mount on confocal microscope equipped with a 40× objective (e.g., LSM 510 Meta DuoScan Spectral Confocal Microscope). Use identical confocal settings for all data collection within the experiment (*see* **Note 7**). Focus using the GFP (488 nm) channel to yield sharp cell surface outlines of the near confluent cells (Fig. 2). There should be about 100 cells per field. Acquire a single optical section using both the GFP and MG (633 nm) channels. Repeat for at least eight separate fields per coverslip.

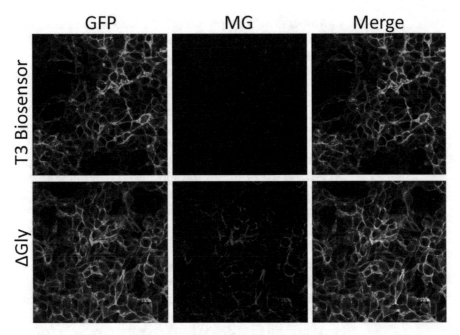

Fig. 2 Fluorescence microscopy data. The figure compares cells expressing the T3 biosensor (*upper row*) to those expressing its activated ΔGly version (*lower row*) in the indicated channels

4. To quantify, for each two-channel image define several regions of interest outside the cells. For each channel in each image, determine the average intensity value in these non-cell areas and use the highest value as the background. For each channel in each image, subtract the background value from all pixels (uniform subtraction). Next, in the GFP channel of each image, use auto threshold to define a threshold and select all above-threshold pixels. Record the average intensity value for these pixels for both the GFP and MG channels. The signal for that image is then expressed as the MG–GFP ratio. Repeat for all eight two-channel images per coverslip. The average of the eight ratios is then the value for a single condition in a single experiment.

5. Comparison of these values between ΔFur, the biosensor, and ΔGly will typically yield a low signal for the biosensor, an even lower signal for ΔFur and a much greater value for ΔGly. To normalize the values for comparison with other sensors subtract the background (ΔFur or untransfected) and then divide by the positive control (ΔGly).

3.2 Assay GalNAc-T Activity Using Biosensor and Multi-well Format Flow-Cytometry

1. Pass the cell lines (ΔFur, biosensor, ΔGly) at 50,000 cells/well into 96-well plates (0.2 ml/well).

2. After 24 h replace the medium with 0.1 ml of the EDTA/PBS solution containing 110 nM MG dye and return to incubator for 5 min (*see* **Note 6**).

3. Suspend cells by pipetting each well up and down 2–3 times with a multichannel pipette with 200 µl tips. Mount plate on flow cytometry instrument (e.g., BD Accuri™ C6 Flow Cytometer). Collect data for 10,000 cells per well using GFP (488 nm) and MG (640 nm) channels (Fig. 3).

4. To quantify, first determine the background. Calculate the geometric mean of values for each channel of each well of the untransfected or ΔFur cells (at least 3 wells). Use the average of these means (one for each channel) as the background values. Next, for each remaining well, determine the geometric mean of the values for each channel. Subtract the channel-specific background values and then calculate the MG–GFP ratio for each well. Averages of these values for multiple trials are then used for comparison.

5. Again, there should be a low signal for the biosensor, an even lower signal for ΔFur (equal to background) and a much greater value for ΔGly. To normalize the values for comparison with other sensors subtract the background (untransfected or ΔFur) and then divide by the positive control (ΔGly).

4   Notes

1. Glycan masking of adjacent protease-processing sites can occur upon addition of a single GalNAc [13, 15] but the current sensors show variable requirements for chain extension. The pan- and T3-specific sensors appear to require extension whereas the T2-specific sensor does not [9, 10].

2. The protocols utilize a membrane impermeant version of the dye and therefore yield surface staining. A membrane permeant version can be used but also yields predominate surface staining

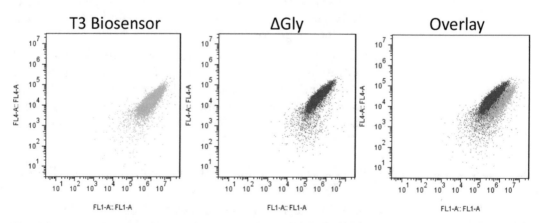

Fig. 3 Flow cytometry data. Shown are scatter plots for the T3 biosensor (*left*), its activated ΔGly version (*middle*), and an overlay. Axes are 488 nm channel (X) and 633 nm channel (Y)

because processing of the biosensor by furin occurs just before its exit from the Golgi. A Golgi signal can be detected if this exit is blocked (e.g., by cell incubation at 20 °C).

3. Excitation and emission spectra for MG11P can be found at http://www.mbic.cmu.edu/images/datasheet/MG-11p-NH2-info_rev21.pdf.

4. Cleavage of the sensor is readily assayed by immunoblot [9]. Briefly, cell lysate proteins are separated by SDS-PAGE, transferred to nitrocellulose and blotted with anti-GFP antibodies to detect uncleaved ($\approx$80 kD) and cleaved ($\approx$60 kD) portions of the sensor. (The missing fragment is secreted into the cell media and detectable with anti-HA antibodies.) Quantify the cleaved fragment as a percent of total. The biosensor should give a value slightly higher than ΔFur and much lower than ΔGly.

5. *Optional linker modification.* The sensor plasmids [9, 10] each contain in-frame fusions encoding a signal sequence, a blocking domain (HL4 heavy chain, [17]), a linker with furin and glycosylation sites, a FAP (L5*, [11]), the myc tag, a transmembrane domain from PDGFR, and either a Venus or GFP domain. The version with the pan-specific glycan acceptor site has a linker with the sequence 5′-c tcg aga aag aag aga tct acc ccc gct cca gct cca tcc ggt ggc ggt ggc agc-3′ encoding NSRKKRSTPAPS where RKKR is the furin site and TPAP comprises the glycan site. To modify selectivity of the sensor, point mutations can be introduced or the entire linker can be substituted with another sequence encoding a glycan acceptor and protease-processing site. The optimal candidates are substrate sequence stretches known to be regulated by glycan masking by a particular GalNAc transferase. In either case, results from published in vitro assays establishing sequence preferences of particular GalNAc transferases [13, 15] can be used to guide further mutagenesis to optimize sensor function [10]. Transient expression can be used for initial tests but ultimately stable cell lines will need to be generated to insure uniform and strong surface expression of the sensors. For any new version, comparison of its signal with and without an intact glycan acceptor site will test its effectiveness. Strong sensors yield maximal activation of 500-fold or more [10]. To confirm isoform specificity of a new sensor, one needs to show that it is strongly activated in cells lacking the particular GalNAc transferase isoform [10]. Genomic editing or RNA interference can be used to suppress expression of the particular isoform.

6. Time after passage can be varied to achieve the desired density. Note that trypsin treatment to passage the cells degrades the surface population of sensor molecules. A minimum of 3 h and a recommended time of at least 6 h post-passage should be used to repopulate the cell surface. Cell release with EDTA/PBS (for passage or analysis) avoids this issue.

7. The intensity of the lasers should be adjusted to ensure maximum sensitivity without saturation. This will depend on expression level, but matched control and experimental sensors should always be analyzed with identical settings.

Acknowledgement

This work was supported by NIH grant GM056779 to A.D.L.

References

1. Moremen KW, Tiemeyer M, Nairn AV (2012) Vertebrate protein glycosylation: diversity, synthesis and function. Nat Rev Mol Cell Biol 13:448–462

2. Schjoldager KT, Clausen H (2012) Site-specific protein O-glycosylation modulates proprotein processing—deciphering specific functions of the large polypeptide GalNAc-transferase gene family. Biochim Biophys Acta 1820:2079–2094

3. van der Post S, Subramani DB, Backstrom M, Johansson ME, Vester-Christensen MB, Mandel U, Bennett EP, Clausen H, Dahlen G, Sroka A, Potempa J, Hansson GC (2013) Site-specific O-glycosylation on the MUC2 mucin protein inhibits cleavage by the Porphyromonas gingivalis secreted cysteine protease (RgpB). J Biol Chem 288:14636–14646

4. Bennett EP, Mandel U, Clausen H, Gerken TA, Fritz TA, Tabak LA (2012) Control of mucin-type O-glycosylation: a classification of the polypeptide GalNAc-transferase gene family. Glycobiology 22:736–756

5. Gill DJ, Tham KM, Chia J, Wang SC, Steentoft C, Clausen H, Bard-Chapeau EA, Bard FA (2013) Initiation of GalNAc-type O-glycosylation in the endoplasmic reticulum promotes cancer cell invasiveness. Proc Natl Acad Sci U S A 110:E3152–E3161

6. Kato K, Jeanneau C, Tarp MA, Benet-Pages A, Lorenz-Depiereux B, Bennett EP, Mandel U, Strom TM, Clausen H (2006) Polypeptide GalNAc-transferase T3 and familial tumoral calcinosis. Secretion of fibroblast growth factor 23 requires O-glycosylation. J Biol Chem 281:18370–18377

7. Kathiresan S, Melander O, Guiducci C, Surti A, Burtt NP, Rieder MJ, Cooper GM, Roos C, Voight BF, Havulinna AS, Wahlstrand B, Hedner T, Corella D, Tai ES, Ordovas JM, Berglund G, Vartiainen E, Jousilahti P, Hedblad B, Taskinen MR, Newton-Cheh C, Salomaa V, Peltonen L, Groop L, Altshuler DM, Orho-Melander M (2008) Six new loci associated with blood low-density lipoprotein cholesterol, high-density lipoprotein cholesterol or triglycerides in humans. Nat Genet 40:189–197

8. Boskovski MT, Yuan S, Pedersen NB, Goth CK, Makova S, Clausen H, Brueckner M, Khokha MK (2013) The heterotaxy gene GALNT11 glycosylates Notch to orchestrate cilia type and laterality. Nature 504:456–459

9. Bachert C, Linstedt AD (2013) A sensor of protein O-glycosylation based on sequential processing in the Golgi apparatus. Traffic 14:47–56

10. Song L, Bachert C, Schjoldager KT, Clausen H, Linstedt AD (2014) Development of isoform-specific sensors of polypeptide GalNAc-transferase activity. J Biol Chem 289:30556–30566

11. Szent-Gyorgyi C, Stanfield RL, Andreko S, Dempsey A, Ahmed M, Capek S, Waggoner AS, Wilson IA, Bruchez MP (2013) Malachite green mediates homodimerization of antibody VL domains to form a fluorescent ternary complex with singular symmetric interfaces. J Mol Biol 425:4595–4613

12. Yoshida A, Suzuki M, Ikenaga H, Takeuchi M (1997) Discovery of the shortest sequence motif for high level mucin-type O-glycosylation. J Biol Chem 272:16884–16888

13. Gerken TA, Jamison O, Perrine CL, Collette JC, Moinova H, Ravi L, Markowitz SD, Shen W, Patel H, Tabak LA (2011) Emerging paradigms for the initiation of mucin-type protein O-glycosylation by the polypeptide GalNAc transferase family of glycosyltransferases. J Biol Chem 286:14493–14507

14. Kong Y, Joshi HJ, Schjoldager KT, Madsen TD, Gerken TA, Vester-Christensen MB, Wandall HH, Bennett EP, Levery SB, Vakhrushev SY, Clausen H (2015) Probing polypeptide GalNAc-transferase isoform substrate specificities by in vitro analysis. Glycobiology 25:55–65

15. Gram Schjoldager KT, Vester-Christensen MB, Goth CK, Petersen TN, Brunak S, Bennett EP, Levery SB, Clausen H (2011) A systematic study of site-specific GalNAc-type O-glycosylation modulating proprotein convertase processing. J Biol Chem 286:40122–40132

16. Schjoldager KT, Vester-Christensen MB, Bennett EP, Levery SB, Schwientek T, Yin W, Blixt O, Clausen H (2010) O-glycosylation modulates proprotein convertase activation of angiopoietin-like protein 3: possible role of polypeptide GalNAc-transferase-2 in regulation of concentrations of plasma lipids. J Biol Chem 285:36293–36303

17. Szent-Gyorgyi C, Schmidt BF, Creeger Y, Fisher GW, Zakel KL, Adler S, Fitzpatrick JA, Woolford CA, Yan Q, Vasilev KV, Berget PB, Bruchez MP, Jarvik JW, Waggoner A (2008) Fluorogen-activating single-chain antibodies for imaging cell surface proteins. Nat Biotechnol 26:235–240

Chapter 11

In Situ Proximity Ligation Assay (PLA) Analysis of Protein Complexes Formed Between Golgi-Resident, Glycosylation-Related Transporters and Transferases in Adherent Mammalian Cell Cultures

Dorota Maszczak-Seneczko, Paulina Sosicka, Teresa Olczak, and Mariusz Olczak

Abstract

In situ proximity ligation assay (PLA) is a novel, revolutionary technique that can be employed to visualize protein complexes in fixed cells and tissues. This approach enables demonstration of close (i.e., up to 40 nm) proximity between any two proteins of interest that can be detected using a pair of specific antibodies that have been raised in distinct species. Primary antibodies bound to the target proteins are subsequently recognized by two PLA probes, i.e., secondary antibodies conjugated with oligonucleotides that anneal when brought into close proximity in the presence of two connector oligonucleotides and a DNA ligase forming a circular DNA molecule. In the next step, the resulting circular DNA is amplified by a rolling circle polymerase. Finally, fluorescent oligonucleotide probes hybridize to complementary fragments of the amplified DNA molecule, forming a typical, spot-like pattern of PLA signal that reflects subcellular localization of protein complexes. Here we describe the use of in situ PLA in adherent cultures of mammalian cells in order to visualize interactions between Golgi-resident, functionally related glycosyltransferases and nucleotide sugar transporters relevant to N-glycan biosynthesis.

Key words In situ proximity ligation assay (PLA), Primary antibodies, Oligonucleotide annealing, DNA amplification, Fluorescence detection, Golgi membrane, Glycosyltransferases, Nucleotide sugar transporters

1 Introduction

The in situ proximity ligation assay (PLA) is a versatile and sensitive technique enabling detection and visualization of close (i.e., up to 40 nm) proximity or interaction between any two biomolecules that can be specifically recognized by primary antibodies. In situ PLA not only demonstrates that an interaction occurs, but—as the name suggests—it also reveals its cellular context. Moreover, since PLA involves the phenomenon of DNA amplification, it

William J. Brown (ed.), *The Golgi Complex: Methods and Protocols*, Methods in Molecular Biology, vol. 1496,
DOI 10.1007/978-1-4939-6463-5_11, © Springer Science+Business Media New York 2016

enables detection of extremely low-abundance protein complexes. PLA was first introduced in 2002 [1] and soon became adapted towards the in situ procedure [2]. Currently, a kit including all reagents (except primary antibodies) necessary to conduct the assay can be purchased from Olink Biosciences (http://www. olink.com) as well as its local representatives.

The first step of the in situ PLA involves dual binding of primary antibodies raised in distinct species to adjacent epitopes that are either exposed on the same protein molecule or contained within two proteins constituting part of a complex. In the second step a pair of species-selective, oligonucleotide-conjugated secondary antibodies, i.e., PLA probes PLUS and MINUS, is applied. This event is followed by the addition of two linear connector oligonucleotides, which use PLA probes as hybridization templates. Upon the addition of a DNA ligase a continuous circular DNA molecule is formed and the oligonucleotide attached to PLA probe PLUS is subsequently used to prime a rolling circle amplification reaction catalyzed by the Phi29 DNA polymerase. This event results in the formation of a single-stranded, concatemeric DNA product that adopts a bundle-like structure being a stable extension of the PLA probe PLUS. The final step involves the addition of fluorophore-labeled oligonucleotide detection probes, which are complementary to multiple sequences generated upon rolling circle amplification and thus can hybridize to them. The successfully conducted assay results in a PLA signal with a characteristic appearance of discrete, bright, fluorescent, submicrometer-size spots that can be easily detected by means of fluorescence microscopy. Each PLA spot corresponds to an individual pair of closely located or interacting proteins revealed by a single rolling circle amplification product labeled with several hundred fluorophore molecules. The outline of the in situ PLA procedure is depicted in Fig. 1 and reflects the 'indirect' detection method (i.e., unlabeled primary antibodies are subsequently recognized by respective PLA probes). 'Direct' detection is also possible to perform, but an effort of conjugating both primary antibodies with respective oligonucleotides has to be made in the first place so that use of PLA probes is not required any more.

In situ PLA was primarily intended to detect close proximity between proteins that are endogenously expressed by cells and tissues [e.g., refs. 3–6]. This strategy appears convenient to conduct, since it requires no genetic manipulations. However, in several cases it cannot be executed. First, there is no primary antibody directed against protein(s) of interest available on the market. Second, it might turn out that the dedicated antibody either does not work properlyintheparticularcelllinemodelusedorisimmunofluorescence-incompatible in general. Finally, cells may produce alternative isoforms (e.g., splice variants) of the same protein that share a common epitope and therefore cannot be distinguished upon binding of an individual primary antibody. The enumerated impediments may be

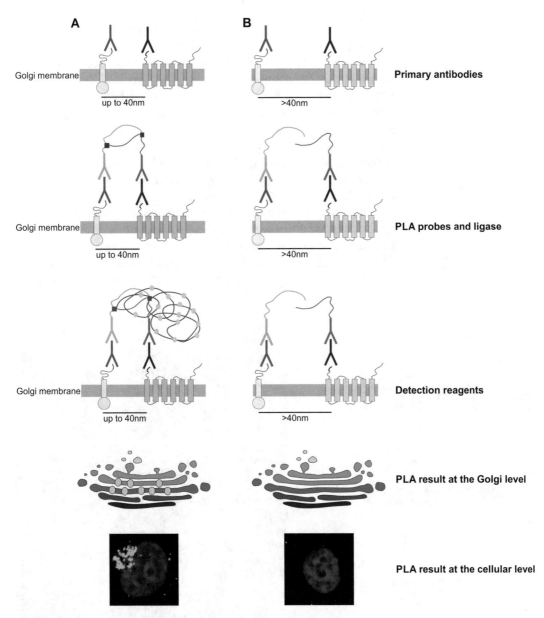

Fig. 1 A schematic, simplified representation of the PLA procedure performed towards a pair of recombinant, Golgi-resident, glycosylation-related transmembrane proteins that have been N-terminally tagged with distinct epitopes (marked in *red* and *dark blue*, respectively). The glycosyltransferase is marked in *yellow*, and the nucleotide sugar transporter is marked in *green*. The results anticipated at different molecular levels for both positive (panel **a**) and negative (panel **b**) combinations are shown. A detailed description of individual PLA steps can be found in the Subheading 1. The two connector oligonucleotides are depicted as *purple squares*, and the PLA signal is represented by a series of *light green circles* arranged along the *rolling circle* amplification product as well as selected Golgi cisternae

136   Dorota Maszczak-Seneczko et al.

readily overcome by in situ PLA detection of transiently expressed, epitope-tagged interaction partners, an approach which was pioneered by Gajadhar and Guha in 2009 [7].

In our hands in situ PLA turned out to be an extremely useful technique for studying protein complexes formed by transmembrane proteins involved in the Golgi-dependent phenomenon of *N*-glycosylation, i.e., specific glycosyltransferases and nucleotide sugar transporters [8]. The hereby presented protocol allows indirect detection of close proximity between these two types of proteins in three main configurations: (1) both interaction partners are expressed exogenously as epitope-tagged fusion proteins (exogenous PLA), (2) the glycosyltransferase is expressed endogenously and the nucleotide sugar transporter is expressed exogenously as an epitope-tagged fusion protein (semi-endogenous PLA), (3) both interaction partners are expressed endogenously (endogenous PLA). Typical results obtained for each of these configurations as well as for the corresponding controls are also presented (*see* Fig. 2).

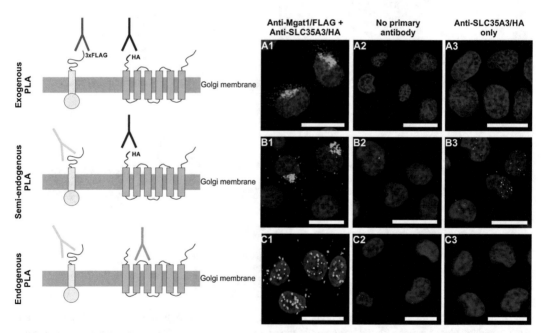

Fig. 2 Representative in situ PLA results obtained for alpha-1,3-mannosyl-glycoprotein 2-beta-*N*-acetylglucosaminyltransferase (Mgat1) and UDP-*N*-acetylglucosamine transporter (SLC35A3). Three PLA configurations are schematically represented in the *left panel* as follows: A, endogenous PLA; B, semi-endogenous PLA; C, exogenous PLA. In endogenous configuration both Mgat1 and SLC35A3 were endogenously expressed. In semi-endogenous configuration Mgat1 was endogenously expressed, while SLC35A3 was exogenously expressed with N-terminally attached HA epitope. In exogenous configuration both Mgat1 and SLC35A3 were exogenously expressed with N-terminally attached FLAG and HA epitopes, respectively. A1-C1, positive samples; A2-C2, negative controls where both primary antibodies were omitted; A3-C3, negative controls where one of the primary antibodies was omitted. Endogenous and semi-endogenous PLAs were performed using PC-3 cells, while exogenous PLA was performed using MDCK-RCA^r cells. Bar, 20 μm

2 Materials

1. An adherent mammalian cell line (e.g., HeLa, HEK293T, COS-7) readily transfectable by routine means (*see* **Note 1**).
2. Culture dishes, multiwell plates, serological pipettes and other cell culture equipment.
3. CO_2 incubator.
4. Complete growth medium (e.g., Dulbecco's Modified Eagle Medium) supplemented with 10 % heat-inactivated fetal bovine serum, 4 mM L-glutamine, 100 IU penicillin and 100 μg/ml streptomycin.
5. 8-well chamber slides (e.g., Millipore, LabTek).
6. 24×60 mm glass coverslips.
7. Phosphate buffered saline (PBS): 50 mM sodium phosphate buffer, 150 mM sodium chloride, pH 7.4.
8. Fixative: 4 % paraformaldehyde in PBS.
9. Permeabilization reagent: 0.1 % Triton X-100 in PBS.
10. Blocking solution and antibody diluent: 10 % normal goat serum in PBS.
11. Primary antibodies specific to Golgi-resident transmembrane proteins of interest (characteristics of all primary antibodies used by us can be found in Table 1).
12. Pipettes covering the range from 1 μl to 1000 μl.
13. Manual repetitive dispenser covering the range from 100 to 200 μl.
14. High purity water.
15. Freeze block for enzymes.

Table 1
Characteristics of recommended primary antibodies

Target protein/epitope	Host	Clonality	Dilution	Vendor
SLC35A2	Rb	PAb	1:50	Abcam
SLC35A3	Rb	PAb	1:50	Abcam
Mgat1	Ms	PAb	1:50	Abcam
Mgat2	Ms	MAb	1:50	Abnova
Mgat4B	Ms	PAb	1:50	Abnova
Mgat5	Ms	MAb	1:20	R&D Systems

Ms mouse, *Rb* rabbit, *PAb* polyclonal antibody, *MAb* monoclonal antibody

16. Vortex.

17. Humidity chamber (*see* **Note 2**).

18. Anti-rabbit PLA probe PLUS (5×) and anti-mouse PLA probe MINUS (5×) (store at 4 °C, *see* **Note 3**).

19. 5× Ligation solution (store at –20 °C).

20. Ligase (1 U/μl, store at –20 °C).

21. 5× Amplification solution (store at –20 °C).

22. Polymerase (10 U/μl, store at –20 °C).

23. Duolink In Situ Wash Buffers A and B: dissolve the entire content of the pouch in 1000 ml of high purity water (store at 4 °C).

24. Duolink In Situ Mounting Medium with DAPI (store at 4 °C).

25. Fluorescence microscope equipped with (a) excitation/emission filters compatible with fluorophore used for detection (*see* **Note 4**) and nuclear stain excitation/emission spectra, (b) 40× and/or 60× oil objectives and (c) image acquisition software.

26. Software for image processing and analysis, e.g., ImageJ (http://rsbweb.nih.gov/ij/) (NIH, Bethesda, MD, USA).

3 Methods

1. Prepare the appropriate amount of 8-well chamber slides depending on the number of samples to be analyzed. Pay attention to include all possible controls (*see* **Note 5**).

2. 24 h prior to the PLA experiment subculture the cells as usual, resuspend the cell pellet in appropriate, complete growth medium and dispense 200–500 μl aliquots into the desired number of wells (*see* **Note 6**). Allow the cells to attach to the slide surface either for a few hours or overnight depending on the adhesive properties of the cell line used.

3. Tap off the growth medium from the slide and wash the slide with PBS (3×5 min) (*see* **Note 7**).

4. Fix the samples with 4% PFA solution in PBS for 20 min at room temperature.

5. Tap off the fixative from the slide.

6. Wash the slide with PBS (3×5 min).

7. Add 0.1% Triton X-100 in PBS to the samples. Incubate the slides for 5 min at room temperature (*see* **Note 8**).

8. Tap off the permeabilization solution from the slide.

9. Wash the slide with PBS (3×5 min).

10. Block nonspecific binding sites with 10% normal goat serum in PBS (*see* **Note 9**). Incubate the slides for 1 h at room temperature.

11. Tap off the blocking solution from the slide.

12. Wash the slide with PBS (3×5 min).

13. Prepare the solution of primary antibodies by diluting with 10 % normal goat serum in PBS (*see* Tables 1 and 2, and **Notes 9–12**).

14. Add the primary antibody solution to the samples. Incubate the slides in a preheated humidity chamber for 2 h at 37 °C (*see* **Note 12**).

15. Tap off the primary antibody solution from the slide.

16. Wash the cells with PBS (3×5 min).

17. Mix and dilute anti-mouse PLA probe MINUS and anti-rabbit PLA probe PLUS 1:5 in 10 % normal goat serum in PBS (*see* **Note 13**). Allow the mixture to stand for 20 min at room temperature.

18. Tap off the primary antibody solution from the slide.

19. Add the PLA probe solution to the samples. Incubate the slides in a preheated humidity chamber for 1 h at 37 °C. Use a mixture of both PLA probes for all control samples as well.

20. Dilute the Ligation stock 1:5 in high purity water and mix.

21. Tap off the PLA probe solution from the slides.

22. Wash the slides with 1× wash buffer A (2×5 min).

23. Remove the Ligase from the freezer using a freezing block (–20 °C). Immediately before addition to the samples add Ligase to the Ligation solution obtained in **step 20** at a 1:40 dilution and vortex.

Table 2
Volumes of individual reagents required to perform in situ PLA procedure using cells cultured in one well of an 8-well chamber slide

Reagent	Volume (µl)
PBS and Duolink in situ wash buffers A and B	200
4 % paraformaldehyde in PBS	150
0.1 % Triton X-100 in PBS	150
10 % normal goat serum in PBS	100
Primary antibody solution	100
PLA probe mixture solution	100
Ligation solution	100
Amplification solution	100

24. Add the Ligation-Ligase solution to each sample. Incubate the slides in a preheated humidity chamber for 30 min at 37 °C.

Detection Reagents are light-sensitive; therefore from this time all incubation and washing steps should be performed in the dark.

25. Dilute the Amplification stock 1:5 in high purity water and mix.

26. Tap off the Ligation-Ligase solution from the slides.

27. Wash the cells with 1× wash buffer A (2 × 2 min).

28. Remove the Polymerase from the freezer using a freezing block (−20 °C). Immediately before addition to the samples add Polymerase to the Amplification solution obtained in **step 25** at a 1:80 dilution and vortex.

29. Add the Amplification-Polymerase solution to each sample. Incubate the slides in a preheated humidity chamber for 100 min at 37 °C.

30. Tap off the Amplification-Polymerase solution from the slides.

31. Wash the slides with 1× wash buffer B (2 × 10 min).

32. Wash the slides with 0.01× wash buffer B (1 × 1 min).

33. Carefully remove the chamber part and gently mount the slide onto a 24×60 mm glass coverslip using Duolink In Situ Mounting Medium with DAPI (*see* **Note 14**). Avoid trapping bubbles between the slide and the coverslip. Seal the edges with nail polish (*see* **Note 15**). Wait for approximately 15 min before proceeding to imaging.

34. View the slide using a confocal microscope. Use exactly the same imaging settings for the samples of interest and controls. Obtain at least 3 images, each showing 3–5 cells. After imaging, store the slides at −20 °C in the dark (*see* **Note 16**).

35. Process the acquired images. Use exactly the same processing mode for the samples of interest and controls. If accurate quantification of the PLA signal is required, analyze the images using the Duolink Image Tool (Olink Biosciences) or equivalent software.

4   Notes

1. Either wild type cells or cells exogenously expressing epitope-tagged fusion protein(s) obtained by means of transfection can be used depending on the availability, performance and mutual compatibility of respective primary antibodies.

2. The samples should not be allowed to dry out at any time before the final step, since it may result in a high background and severe artifacts. Therefore, it is critically important that sufficient humidity is ensured during all incubation steps.

This can be achieved by performing all incubation steps in a humidity chamber (e.g., a tightly covered small box with moist paper towels in the bottom). Moreover, we strongly recommend keeping a moist sheet of paper towel underneath the cover of the chamber slide during all incubation steps.

3. Anti-rabbit PLA probe PLUS is a secondary antibody conjugated to oligonucleotide PLUS, designed to bind in a specific manner to any rabbit-derived primary antibody. Anti-mouse PLA probe MINUS is a secondary antibody conjugated to oligonucleotide MINUS, designed to bind in a specific manner to any mouse-derived primary antibody. An inverse combination, i.e., anti-rabbit PLA probe MINUS and anti-mouse PLA probe PLUS can be as well used to detect a pair of rabbit and mouse primary antibodies. The key issue is that PLA probe PLUS must always be combined with PLA probe MINUS.

4. Depending on the light sources and filters available on the microscope, four spectrally different Duolink In Situ Detection Reagents can be used for fluorescent PLA detection (Green, Orange, Red and FarRed).

5. A positive PLA result can only be considered reliable if all appropriate controls have been included in the same experiment (i.e., performed in parallel by applying exactly the same conditions to all the samples analyzed). Ideally, three different types of controls should be performed. First, a positive control should be included to verify that the PLA procedure has been properly conducted. It should involve a pair of primary antibodies targeting epitopes that are known to reside in close proximity (e.g., two distinct proteins or two distinct epitopes present within the same protein molecule). Moreover, two types of negative controls should be performed. The first type, called the biological control, should include cells that do not express one or both proteins of interest. While this is relatively easy to achieve when exogenously expressed, epitope-tagged proteins are being detected (non-transfected cells may simply be employed), such a control may be barely accessible when endogenously expressed proteins are being investigated unless a corresponding knockout cell line is available. This type of control allows one to validate the specificity of the primary antibodies used. Finally, a series of technical negative controls should be performed as follows: (a) the one primary antibody of interest is omitted, (b) the other primary antibody of interest is omitted, (c) both primary antibodies of interest are omitted. These controls are designed to reveal the background triggered by the PLA probes themselves in the particular system used.

6. For the majority of cell lines seeding between 30,000 and 40,000 cells per well should result in 70–90% confluency on the following day, which is optimal for a PLA experiment.

7. Use of a manual repetitive dispenser allows one to save time during washing between consecutive incubation steps. It may also be used to dispense reagent solutions.

8. Efficient permeabilization is critically important when immunostaining of Golgi-resident transmembrane proteins, such as glycosyltransferases, is being performed, since the corresponding epitopes are often luminally oriented.

9. It is of crucial importance to use a proper blocking solution and antibody diluent. Follow the recommendations from the vendor of your primary antibodies, if available. If you have previously optimized your assay in immunofluorescence, use exactly the same conditions for the in situ PLA procedure. The hereby suggested antibodies are compatible with 10% normal goat serum in PBS used as the blocking solution as well as antibody diluent. However, a blocking solution and antibody diluent is included in the PLA probes in a ready-to-use concentration for assays involving primary antibodies that have not been optimized in these terms.

10. When performing indirect detection, a combination of primary antibodies raised in different species must always be used so they can be distinguished by respective PLA probes. Hereby suggested antibodies are of either rabbit or mouse origin. However, anti-goat PLA probes PLUS and MINUS are also available when required. Alternatively, one can perform direct detection using primary antibodies raised in the same species that have been previously labeled with PLA oligonucleotide arms PLUS and MINUS. Direct detection has one major advantage, namely a decrease in the PLA working distance due to elimination of an entire layer of reagents, i.e., PLA probes.

11. A successful result of the PLA procedure strictly relies on the performance of primary antibodies, which should be of IgG class, specific for the target to be detected and preferably affinity purified. Either polyclonal or monoclonal antibodies can be used. If possible, antibodies that are IHC and/or IF classified should be chosen. Otherwise, it is strongly recommended to validate each antibody by performing a regular immunofluorescent staining using the same cell line that is intended to be used in the PLA procedure. Both antibodies should bind to the target epitopes under the same conditions (fixation, permeabilization, blocking reagent, washing buffer, etc.).

12. It is crucial for the performance of the assay to optimize the conditions for the primary antibodies. These conditions should be optimized for each cell line used as well, since distinct cell lines may exhibit significantly different levels of the target proteins.

13. The buffer conditions for the secondary PLA probes should be exactly the same as those for the primary antibodies to avoid

background staining. A diluent for primary antibodies and PLA probes should also contain the same agent which was used for blocking of nonspecific binding sites.

14. It is highly advisable to use Duolink In Situ Mounting Medium with DAPI for the final preparation of PLA-analyzed slides, since it is optimal for preservation of the PLA signal. Using other mounting media may cause several problems including an insufficient PLA signal or red appearance of nuclei upon fluorescence excitation.

15. It is important to remember that Duolink In Situ Mounting Medium with DAPI does not solidify.

16. To our best knowledge and experience, slides stored at −20 °C in the dark for at least several months should not display any change in the PLA signal intensity. It is extremely important to use Duolink In Situ Mounting Medium with DAPI in order to preserve PLA signals if slides are to be stored for extended periods of time.

Acknowledgement

This work was supported by grant number 2014/15/B/NZ3/00372 from the National Science Center (NCN, Krakow, Poland).

References

1. Fredriksson S, Gullberg M, Jarvius J et al (2002) Protein detection using proximity-dependent DNA ligation assays. Nat Biotechnol 20:473–477

2. Soderberg O, Gullberg M, Jarvius M et al (2006) Direct observation of individual endogenous protein complexes *in situ* by proximity ligation. Nat Methods 3:995–1000

3. Trifilieff P, Rives M-L, Urizar E et al (2011) Detection of antigen interactions *ex vivo* by proximity ligation assay: endogenous dopamine D2-adenosine A2A receptor complexes in the striatum. Biotechniques 51:111–118

4. Bellucci A, Navarria L, Falarti E et al (2011) Redistribution of DAT/a-synuclein complexes visualized by "*in situ*" proximity ligation assay in transgenic mice modelling early Parkinson's disease. PLoS One 6, e27959

5. Hägglund MGA, Hellsten SV, Bagchi S et al (2013) Characterization of the transporter B0AT3 (Slc6a17) in the rodent central nervous system. BMC Neurosci 14:54

6. Bart G, Vico NO, Hassinen A et al (2015) Fluorescence resonance energy transfer (FRET) and proximity ligation assays reveal functionally relevant homo- and heteromeric complexes among hyaluronan synthases HAS1, HAS2, and HAS3. J Biol Chem 290:11479–11490

7. Gajadhar A, Guha A (2009) A proximity ligation assay using transiently transfected, epitope-tagged proteins: application for *in situ* detection of dimerized receptor tyrosine kinases. Biotechniques 48:145–152

8. Maszczak-Seneczko D, Sosicka P, Kaczmarek B et al (2015) UDP-galactose (SLC35A2) and UDP-N-acetylglucosamine (SLC35A3) transporters form glycosylation-related complexes with mannoside acetylglucosaminyltransferases (Mgats). J Biol Chem 290:15475–15486

Chapter 12

Creating Knockouts of Conserved Oligomeric Golgi Complex Subunits Using CRISPR-Mediated Gene Editing Paired with a Selection Strategy Based on Glycosylation Defects Associated with Impaired COG Complex Function

Jessica Bailey Blackburn and Vladimir V. Lupashin

Abstract

The conserved oligomeric Golgi (COG) complex is a key evolutionarily conserved multisubunit protein machinery that regulates tethering and fusion of intra-Golgi transport vesicles. The Golgi apparatus specifically promotes sorting and complex glycosylation of glycoconjugates. Without proper glycosylation and processing, proteins and lipids will be mislocalized and/or have impaired function. The Golgi glycosylation machinery is kept in homeostasis by a careful balance of anterograde and retrograde trafficking to ensure proper localization of the glycosylation enzymes and their substrates. This balance, like other steps of membrane trafficking, is maintained by vesicle trafficking machinery that includes COPI vesicular coat proteins, SNAREs, Rabs, and both coiled-coil and multi-subunit vesicular tethers. The COG complex interacts with other membrane trafficking components and is essential for proper localization of Golgi glycosylation machinery. Here we describe using CRISPR-mediated gene editing coupled with a phenotype-based selection strategy directly linked to the COG complex's role in glycosylation homeostasis to obtain COG complex subunit knockouts (KOs). This has resulted in clonal KOs for each COG subunit in HEK293T cells and gives the ability to further probe the role of the COG complex in Golgi homeostasis.

Key words CRISPR, COG, Conserved oligomeric Golgi complex, Fluorescently tagged lectins, FAC sorting, Glycosylation defects, Knockouts, SubAB toxin, Cholera toxin

1 Introduction

The COG complex is a bi-lobed protein complex that has eight subunits (lobe A: 1–4 lobe B: 5–8) and functions in retrograde trafficking of vesicles (containing glycosylation enzymes among other cargo) at the Golgi apparatus [1–6]. Its conservation across a variety of organisms (yeast, worms, plants, and humans to name a few) [1, 7–11] hints at its crucial role in the cell, and indeed a loss of function of the COG complex results in a genetic disorder in the family of Congenital Disorders of Glycosylation (CDG) [12]. These COG associated disorders fall under type II which is

William J. Brown (ed.), *The Golgi Complex: Methods and Protocols*, Methods in Molecular Biology, vol. 1496, DOI 10.1007/978-1-4939-6463-5_12, © Springer Science+Business Media New York 2016

characterized by the improper processing of glycosylated proteins, presumably due to lack of proper glycosylation enzyme homeostasis. Patients with these disorders exhibit a wide variety of symptoms due to the many proteins that are dependent upon COG for proper location and function. Some of these include; issues with clotting, elevated liver enzymes, mental and growth retardation, seizures, and—in the most severe cases—lethality within the first year of life [13, 14]. Additionally, pathogens, such as *Chlamydia* and HIV, are known to depend on COG function for their trafficking and survival in host cells [15–17]. Though it is clear that the COG complex is important for Golgi homeostasis, its specific mechanism of action is still being elucidated.

One of the main drawbacks of COG complex studies has been a lack of immortal human cell lines that are deficient in each COG subunit to provide a clean background for functional assays. The current work-around for this is to use RNAi approaches to significantly reduce protein levels [18–20], or to re-localize over expressed COG protein to a non-physiological location using a mitochondrial targeting tag [3].

The following method alleviates this issue by creating immortal human KO cell lines deficient for individual COG complex subunits using CRISPR technology paired with a COG deficiency specific sorting methodology we developed for this project. The phenotypic differences between the KOs and their parental cell lines are more numerous and far more drastic than seen in the past knockdown and re-localization assays used to characterize the COG complex thus far.

CRISPR-Cas9 gene editing has made a great impact in many fields since its discovery as a gene editing platform in 2012 [21]. The CRISPR-Cas9 system allows for quick gene KO or modification in a highly specific manner by using a single guide RNA sequence, or sgRNA, of 17–20 base pairs to target bacterial endonuclease Cas9 to create a double strand break in the genomic DNA. This sgRNA can be any sequence as long as it lies adjacent to an NGG protospacer motif (or PAM) sequence. The double strand endonuclease action causes a homozygous KO <1% to 99% of the time depending on transfection conditions.

Because of the variable efficiency, CRISPR-Cas9 treated cells must be screened to determine that the gene of interest has been homozygously knocked-out.

In this chapter we detail a gene knockout and selection method based on the arising glycosylation defects resulting from COG complex malfunction that have previously been shown [22].

We also include a section that highlights two examples of phenotypic analysis that have been done with the resulting KO cells.

2 Materials

All buffers should be prepared using ultrapure water (18 MΩ water) except for electrophoresis (running) and electroblot buffers. All reagents are analytical grade.

2.1 Transfection of Cells with CRISPR Constructs

1. *E. coli* NEB 10-beta Competent *E. coli* for transformation and obtaining plasmids (*see* **Note 1**).

2. SOC (Super Optimal Broth) media.

3. LB (Luria–Bertani) broth: add 20 g of LB Broth to 1 L of water and autoclave for 30 min.

4. Kanamycin antibiotic (1000× stock: 150 mg kanamycin sulfate in 5 mL dd H_2O).

5. Kanamycin (30 mg/L) LB agar plates.

6. Plasmids containing Cas9 endonuclease and gRNA targeting the first exon of the COG 7 subunit (Catalog # HCP222287-CG01-3-B-a, HCP222287-CG01-3-B-b, HCP222287-CG01-3-B-c) (Gencopoeia, Rockville, MD).

7. Control plasmid containing GFP (*see* **Note 2**).

8. QIAprep Spin Miniprep Kit.

9. HEK293T cells (CRL-3216 ATCC, Mananas, VA) grown to ~70% confluency on 6-well plates.

10. Lipofectamine 2000 Transfection Reagent.

11. Dulbecco's Phosphate Buffered Saline (DPBS 1×) without calcium and magnesium.

12. Transfection media: Opti-MEM I Reduced Serum Media buffered with HEPES and sodium bicarbonate and supplemented with hypoxanthine, thymidine, sodium bicarbonate and supplemented with hypoxanthine, thymidine, sodium pyruvate, L-glutamine, trace elements, and growth factors.

13. Growth media: dilute 50 mL of heat-inactivated Fetal Bovine Serum (FBS) in 450 mL of DMEM/F-12 50/50 medium supplemented with 2.5 mM L-glutamine, 15 mM HEPES. Sterile filtered.

2.2 Lectin Staining, FAC Sorting, and Colony Expanding

1. 12 mm round glass coverslips (#1.5, 0.17 mm thickness).

2. Glass slides: frosted microscope slides (precleaned).

3. *Galanthus nivalis* lectin (GNL) (20 µg/mL, Vector Laboratories; Burlingame, CA) labeled with an Alexa 647 protein labeling kit (Life Technologies) according to the manufacturer protocol.

4. 0.1% bovine serum albumin (BSA) in PBS, 0.2 µm sterile filtered.

5. Collagen solution: 50 µg/mL Collagen, Bovine, Type I in 0.01 N HCl.

6. Paraformaldehyde 16% stock solution.

7. Prolong® Gold antifade reagent.

8. Cell Sorting Media: PBS, 25 mM Hepes pH = 7.0, 2% FBS (heat inactivated) 1 mM EDTA, 0.2 µm sterile filtered.

9. Mammalian cell Culture Plates: 10 cm, 12-well, 24-well, and 96-well.

10. 1.5 mL microcentrifuge tubes.

11. Filtered Cap 5 mL—12 × 75 mm polystyrene round bottom tubes.

12. Gibco™ 100× Antibiotic/Antimycotic.

13. Widefield or confocal fluorescence microscope.

2.3 Lysate and DNA Preparation for Validation

1. 2% SDS in water.

2. Heat blocks set to 95 °C for lysates and 63 °C and 98 °C for DNA.

3. 1.5 mL microcentrifuge tubes.

4. Quick Extract DNA extraction solution.

2.4 Sodium Dodecyl Sulfate– Polyacrylamide Gel Electrophoresis and Western Blot Components

1. 9% SDS-PAGE mini-gels.

2. Vertical mini-gel electrophoresis unit.

3. SDS PAGE running buffer: 25 mM Tris, 0.192 M glycine, 0.1% SDS.

4. 0.2 µm Whatman Protran Nitrocellulose Blotting Membranes.

5. Blotting paper.

6. Pierce G2 Fast Blotter and 1-Step™ Transfer Buffer (Thermo Scientific, Rockford, IL).

7. Odyssey™ blocking buffer (LI-COR Biosciences, Pittsburgh, NE).

8. Secondary antibody incubation solution: PBS containing 5% nonfat dry milk.

9. Antibodies:

    (a) Primary: rabbit anti-COG7 antibody were developed using purified His6-COG7 expressed in bacteria [2] and mouse monoclonal β-actin (Sigma #030 M4788).

    (b) Secondary: LI-COR Donkey anti-Mouse 680 (#926-68072), LI-COR Goat anti-Rabbit 800 (#925-32211) (LI-COR Biosciences, Lincoln England).

10. LI-COR Odyssey™ imaging system.

11. 6× Laemmli sample buffer with 5% 2-mercaptoethanol.

12. Ponceau S stain: 0.2 g Ponceau S, 5 mL of Acetic acid (100%), 200 mL dd H$_2$O.

13. A Western Incubation Box.

14. Dulbecco's Phosphate Buffered Saline (DPBS 1×) without calcium and magnesium.

15. Fat free evaporated milk.

2.5 Polymerase Chain Reaction (PCR) and Purification for Sequencing

1. Primers used to amplify and targeting region of COG 7 gene:

   (a) COG 7 Forward Primer: AGAGGAGGAAAAACAACAC CCAA.

   (b) COG 7 Reverse Primer: AGTTACCCGTCCTGGCGTTT.

2. DMSO.

3. dNTPs.

4. 10× exTaq buffer.

5. exTaq polymerase.

6. 0.2 mL PCR tubes.

7. Thermocycler.

8. Zymoclean gel DNA recovery kit.

2.6 Phenotypic Analysis by SubAB Trafficking

1. 6-well plates (TPP) with cells between 50% and 80%.

2. SubAB toxin [22].

3. Growth media (*see* Subheading 2.1, **item 13**).

4. 95 °C 2% SDS.

5. Antibodies:

   (a) Primary: goat polyclonal anti-GRP78 (Santa Cruz Biotechnology, Santa Cruz, CA, SC-1051) and mouse monoclonal β-actin (Sigma #030M4788).

   (b) Secondary: Donkey anti-Mouse 680, Donkey anti-Goat 800 (LI-COR Biosciences).

6. Gradient gels 4–15% (commercially available).

7. All materials listed in Subheading 2.4 for SDS-PAGE and Western blotting.

2.7 Phenotypic Analysis by Cholera Toxin Binding and Flow Cytometry

1. 6-well plates with cells (COG KO and WT HEK293T cells as a control) with confluency between 50% and 80%.

2. Cholera toxin lectin (B subunit only) (CTxB) (20 μg/mL) was labeled with an Alexa 647 protein labeling kit from Life Technologies.

3. Growth media (*see* Subheading 2.1, **item 13**).

4. 0.1% BSA in PBS. 0.2 μm sterile filtered.

5. 1.5 mL microcentrifuge tubes.

6. Filtered Cap 5 mL—12×75 mm polystyrene round bottom tubes.

7. 7.1 μM 4′,6-diamidino-2-phenylindole, dihydrochloride (DAPI).

8. BD Fortessa Flow Cytometer (or equivalent) detecting at 647 nm.

3 Methods

3.1 Plasmid Transformation and Preparation

Day One

1. Add 1 μL of COG 7 targeting CRISPR plasmid to the tube containing competent *E. coli* cells; let incubate for 30 min on ice.

2. Heat-shock the cells: Place tube with cells in a 42 °C water bath for 40 s.

3. Put tube back on ice for 2 min for recovery.

4. Add 300 μL SOC media to cells.

5. Incubate in 37 °C water bath for 1 h.

6. Spin down for 1 min (6000 xg).

7. Remove 200 μL and resuspend pellet in remaining liquid.

8. Take suspension and spread on a LB Agar plate (with kanamycin, encoded for by the CRISPR plasmid).

9. Place plate in incubator at 37 °C overnight.

Day Two

10. Pick up a single colony with sterile plastic loop and place it in 5 mL of LB broth (with kanamycin, encoded for by the CRISPR plasmid).

11. Let bacteria incubate in a shaker set at 200 rpm and 37 °C overnight.

12. Overnight culture was then used to obtain DNA using the QIAprep Spin Miniprep Kit, following the manufacturer's instructions.

3.2 Transfection

1. Plate HEK293T cells on a 6-well plate so that they will be at 50–70 % confluency the next day, approximately 1.5×10^6 cells/well.

2. The following day transfect the cells using mini prepped DNA and following the Lipofectamine 2000 standard protocol (*see* **Note 3**).

3. Remove OptiMem containing Lipofectamine 2000 and DNA the next morning and replace with regular culture media (*see* **Note 4**).

3.3 *Galanthus nivalis* Lectin Labeling

Lectin staining (*see* **Note 5**) is performed with both fixed and unfixed cells as in [23, 24] with some modifications. Solutions are stored at 4 °C until use.

1. 5–7 days after transfection, remove the 6-well plate containing cells transfected with CRISPR plasmids from the incubator. Resuspend the cells by gently pipetting up and down and take an aliquot of cells (*see* **Notes 6** and **7**) to plate in a 24-well plate on collagen coated coverslips (*see* **Note 8**) (Keep a large portion on the 6-well plate for cell sorting).

2. 24–48 h after plating cells prepare to do lectin labelling and immunofluoroscence.

3. Wash cells grown on coverslips 3 times with PBS (*see* **Note 9**).

4. Fix each coverslip for 15 min with 1 % PFA in PBS (made from 16 % stock solution).

5. After fixation, incubate cells with 1 % BSA in PBS for 10 min to block nonspecific binding. Repeat.

6. Incubate cells with *Galanthus nivalis* lectin (GNL) labeled with Alexa Fluor 647 diluted 1:500 in 1 % BSA in PBS for 30 min. (Keep cells in the dark during incubation)

7. Wash cells 5 times with DPBS and then fix again with 4 % PFA for 15 min.

8. Wash cells 5 times with PBS.

9. Wash with 0.1 % Triton-X 100 in PBS for 1 min for further permeabilization.

10. Dunk coverslips in PBS containing DAPI 10 times then dunk coverslips in water 10 times before mounting on glass microscope slides using Prolong® Gold antifade reagent.

11. Coverslips are then cured in the dark overnight at room temperature.

12. Examine cells and obtain images with fluorescence microscope. Our images were taken on a Zeiss LSM510 confocal microscope with a 63× oil immersion objective.

3.4 FAC Sorting

1. After ensuring that a population of your cells are GNL positive via immunofluorescence, remove the remaining CRISPR transfected cells from the 6-well plate via resuspension in regular culture media by gentle pipetting up and down with 1 mL pipette, then place in 1.5 mL microcentrifuge tubes and spin down at $600 \times g$ for 3 min.

2. Remove the media carefully and resuspend cells in ice-cold 0.1 % filtered BSA in PBS solution.

3. Spin down cells again then resuspend in ice-cold 0.1 % BSA solution containing GNL labeled with Alexa Fluor 647 at a 1:1000 dilution, and place on ice in the dark for 30 min.

4. Spin cells down ($600 \times g$ for 3 min), wash three times with 0.1 % BSA solution, resuspend in ice-cold cell sorting media (sterile filtered), and pipet through filtered Cap 5 mL—12×75 mm polystyrene round bottom tubes to remove clumps. Place tubes on ice.

5. Cells need to be sorted with a cell sorting capable flow cytometer that can properly register 647 nm. We use the FACSAria (BD Biosciences) and run wild type GNL-647 stained and unstained cells as a control each time.

6. The FACSAria can sort one lectin positive cell per well in a 96-well plate. 100 μL of culture medium containing 1× Gibco Antibiotic/Antimycotic is added into each well prior to sorting (*see* **Notes 10** and **11**).

3.5 Expanding Colonies and Lectin Testing

1. 10 days after sorting, the 96-well plates were screened for growing patches of cells (*see* **Note 12**).

2. Mark wells which only have one patch of cells growing as they are assumed to be from one common cell.

3. 15 days after sorting split wells that have colonies growing and grow up in 24-well plates (*see* **Note 13**) with collagen coated coverslips for lectin staining analysis as detailed above (Fig. 1a).

3.6 Validation Using SDS-PAGE and Western Blot

1. After colonies are confirmed to be positive for increased GNL binding, colonies were expanded onto one well of a 6-well plate and grown to ~90 % confluency.

2. Resuspend cells in 1 mL of regular culture medium and take a 100 μL aliquot for cell counting for chromosomal DNA prep (*see* **Note 14**). Include a wild type cell sample as well for a control.

3. After 100,000 cells are removed for chromosomal DNA prep (*see* Subheading 3.7 for more details), move the remaining cells to a 1.5 mL microcentrifuge tube.

4. Spin cells down at $600 \times g$ for 3 min.

5. While cells are spinning down, heat 2 % SDS to 95 °C.

6. Remove the supernatant from the cells and resuspend in PBS. Spin down again.

7. Remove supernatant again and add 250 μL of 95 °C 2 % SDS to cells then resuspend and heat at 95 °C for 5 additional min.

8. Add 50 μL of 6× sample buffer to cells and vortex.

9. Heat solution for 3 additional minutes at 95 °C then remove from heat (*see* **Note 15**).

10. Store lysate at –20 °C until needed.

11. Load 10 μL of lysate per well in a 9 % SDS-PAGE gel and run at 180 V until the dye front reaches the bottom of the gel. Turn off the machine.

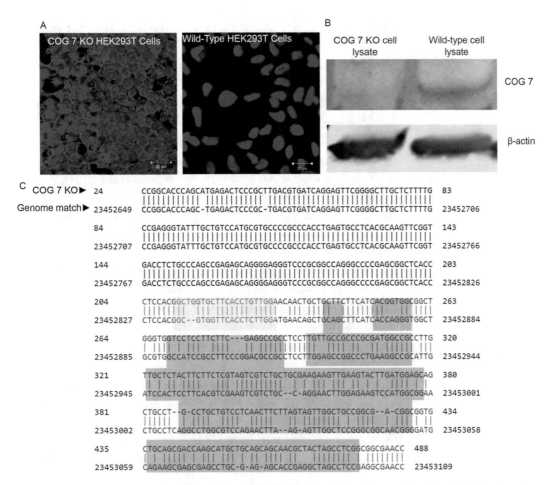

Fig. 1 Analysis of COG 7 KO: (**a**) Lectin staining of WT HEK293 and COG 7 KO HEK293 cells with *Galanthus niva-lis*lectin-Alexa 647 (GNL). Cells were mounted in DAPI containing media to visual the nuclei (*blue*). GNL is false colored *pink*. Images were takes with a 63× immersion lens on a confocal microscope. Settings were calibrated using stained control cells and kept the same throughout. (**b**) Western immunoblot for WT HEK and COG 7 KO cells: delta 7 cell line blotted next to WT HEK293T lysate as a control. (**c**) Sequencing and BLAST of COG gene in KO cells: PCR products were obtained by amplifying the gene of interest from chromosomal DNA. These products were then PCR cleaned and sent for sequencing. Products appeared to be of the expected size by gel (531 bp). Once the sequence was obtained this was BLASTed against the human genome. The expected cut site based on the guide RNA is highlighted in *yellow* and resultant frameshift mutations can be seen in *red*. WT HEK293 cells were also sequenced and BLASTed as a control. Frameshift mutations were not seen in control samples

12. Soak 2 thick sheets of filter paper and the nitrocellulose membrane in 1-Step™ Transfer Buffer.

13. Remove the gel from the chamber and pry open glass. Place gel in ddH₂O.

14. Place one piece of filter paper soaked in 1-Step™ Transfer Buffer on the anode side of the Pierce G2 Fast Blotter cassette. Then place soaked nitrocellulose membrane on top of this.

15. Next, carefully place gel on nitrocellulose membrane and roll gel with clean roller to remove air bubbles.

16. Place second soaked sheet of filter paper on top of the gel and carefully roll again to remove any residual air bubbles.

17. Place cathode portion on top and press to secure the cassette. Place cassette into Fast Blotter machine.

18. Select Mixed Range MW protocol and press Start.

19. After transfer, stain the membrane with Ponceau S stain for 5 min to ensure the transfer went well and also to assess over-all protein levels in each lysate.

20. Trim down membrane and place in a blotting box with PBS. Rinse membrane in PBS (3× for 4 min each) at room temperature while rocking on a table rocker.

21. Block the membrane for 20 min using Odyssey blocking buffer while rocking at room temperature.

22. Add primary antibodies and incubate overnight, on a table rocker at 4 °C.

23. Wash the membrane 3× for 4 min each with PBS while rocking.

24. Add the secondary antibody solution diluted in PBS containing 5 % milk. Incubate in the dark at room temperature while rocking for 40 min.

25. Repeat washing step (**step 23**).

26. Scan blot on LI-COR Odyessy™ imaging system (Fig. 1b).

3.7 Validation Using PCR and Sequencing

1. After colonies were confirmed to be positive for GNL binding, expand each colony onto one well of a 6-well plate and grown to ~90 %.

2. Resuspend cells from each colony in 1 mL of regular culture medium take a 100 µL aliquot for cell counting. Include a wild type cell sample as well for a control.

3. After determining number of viable cells per µL, remove 100,000 cells and place in a microcentrifuge tube.

4. Spin down cells at $600 \times g$ for 3 min, remove supernatant.

5. Add 0.5 mL of Quick Extract DNA Solution from Epicenter, using the recommended protocol for cells (*see* **Note 16**).

6. Store at −20 °C until needed.

7. For chromosomal PCR use the following for 50 µL reactions:
   2 µL of 10 mM each of Cog 7 forward and reverse primer
   2–3 µL of the chromosomal DNA
   2.5 µL of DMSO
   4 µL of dNTPs
   5 µL of 5× exTaq buffer

0.2 µL of exTaq polymerase
*Do two reactions for each colony.

8. Place tubes in a thermocycler and use the following run settings (*see* **Note 17**):

    (a) 95 °C, 1:00 min

    (b) 95 °C, 0:10

    (c) 57 °C, 0:10

    (d) 72 °C, 0:30

    (e) Repeat (b)–(d), 35×

    (f) 72 °C, 0:30

    (g) 4 °C, ∞

9. Add loading dye to each reaction then load on a 1% agarose gel

10. Purify product of correct size using the Zymoclean gel DNA recovery kit and following the manufacturer's instructions (*see* **Note 18**).

11. After eluting, take 2 µL of gel purified product and dilute in water with loading dye. Run on a gel to ensure a product of the correct size was obtained.

12. Sequence samples. (We then send samples for sequencing at the UAMS core sequencing facility.)

13. Run the resulting sequences against the human genome using NCBI BLAST™ to look for mismatches, deletions and insertions near CRISPR cut site (Fig. 1c).

14. Optional further verification (*see* **Note 19**).

3.8 Phenotypic Analysis Using SubAB Trafficking

Analysis of Subtilase cytotoxin (SubAB) trafficking was performed as in [22] with some modifications. In brief, cells are incubated with the toxin for varying lengths of time. Lysates are taken after each time point, then analyzed for GRP78 cleavage via western blot. The amount of time it takes for SubAB to cleave its target (GRP78) (determined by western blot) is an indicator for the efficiency of retrograde trafficking from the plasma membrane through the Golgi and to the ER. We have shown previously that SubAB retrograde trafficking is delayed in COG complex deficient cells [22].

1. Grow each KO clone on 6-wells of a 6-well plate. Also grow wild-type HEK293T cells.

2. Each well will be a different time point. Label the wells: 0 min, 20 min, 40 min, 60 min, 120 min, 180 min. Remove media from each well and replace with 500 µL of fresh media. Place back at 37 °C.

3. Make a 6× stock of SubAB toxin in culture medium. At each time point 100 µL will be added to the appropriate well to get a final working concentration of 20 µg per mL (*see* **Note 20**).

4. Add 100 μL of 6× SubAB stock to the wells labeled 180 min. Your experiment will end 3 h from this time point. Place back at 37 °C.

5. Continue adding toxin to the wells at the appropriate time points with relation to the end time. Gently rock the plate back and forth to mix the toxin with the media. Place back at 37 °C after each addition of toxin until the next time point is reached.

6. After 3 h from the initial toxin addition, remove media containing SubAB from all wells and place this into a sealed tube and throw into a biohazard bag. Replace with 1 mL of PBS.

7. Resuspend cells by gently pipetting up and down. Place cell suspension into a microcentrifuge tube. Each time point of each sample goes into a separate tube.

8. Spin cells down at $600 \times g$ for 3 min.

9. Remove PBS and lyse samples with 250 μL of 95 °C 2 % SDS (*see* **Note 21**). Vortex, then heat samples for 5 min at 95 °C.

10. Follow **steps 8–26** of Subheading 3.6 for running the gel and blotting (Fig. 2a).

3.9 Phenotypic Analysis Using Cholera Toxin

Cholera Toxin binds to GM1 ganglioside residues. These glycolipids have a terminal sialic acid. Sialic acid residues are added in the trans Golgi, so only cells with properly functioning trans Golgi glycosylation machinery will have these modifications. COG complex deficient cells demonstrate altered sialylation of plasma membrane glycoconjugates [23]. This assay is a complement to the GNL screening (high GNL binding and low Cholera Toxin binding in COG KO cells).

1. Grow each KO clone (and wild-type HEK293T cells as a control) to ~90 % confluency on a 6-well plate. Have one well of a 6-well plate for each cell clone.

2. Remove media and resuspend cells in 1 mL of PBS. Place in 1.5 mL microcentrifuge tubes and spin down at $600 \times g$ for 3 min.

3. Remove the PBS carefully and resuspend cells in ice-cold 0.1 % filtered BSA in PBS solution.

4. Spin down cells again at $600 \times g$ for 3 min, then resuspend in ice-cold 0.1 % BSA solution containing Cholera Toxin (*see* **Note 22**) labeled with Alexa Fluor 647 at a 1:1000 dilution and place tubes on ice in the dark for 30 min.

5. After this, spin cells down again at $600 \times g$ for 3 min then resuspended in 500 μL of ice-cold 0.1 % BSA solution DAPI (1:10,000) and placed into a 5 mL—12 × 75 mm polystyrene round bottom tube. Pipet up and down to remove clumps. Place tubes on ice. (*see* **Note 23**)

6. Cells need to be analyzed with a flow cytometer that ca that can properly register Alexa 647. We use the BD Fortessa at the

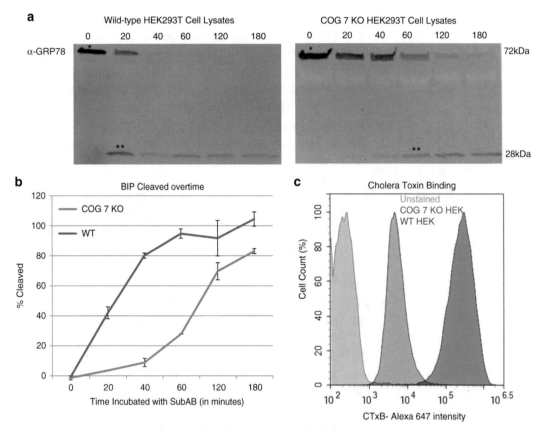

Fig. 2 Phenotypic changes in COG 7 KOs: (**a**) SubAB toxin trafficking and activity is affected in COG 7 KO cells: Cells were incubated at the indicated times (in minutes) with SubAB toxin. Cells were collected and lysates were made as described in Subheading 3.8. *Asterisk* indicates full length GRP78 and *asterisks* indicates cleaved GRP78, mediated by SubAB after it has reached the ER. 50 % cleavage can be seen in 20 min in control cells, but it takes 60–120 min for the same amount of cleavage in COG 7 KO cells (**b**) Quantification of GRP78 cleavage over time. Two separate COG 7 KOs were run. (**c**) Cholera Toxin binding is affected in COG 7 KO cells: Cells were incubated with CTx-647 for 30 min on ice flow cytometry was performed to analyze CTx binding as described in Subheading 3.9. COG 7 KO cells have a lesser binding capacity than WT cells, possibly due to mis-glycosylated GM1, the target of CTx binding

UAMS Flow Cytometry Core and run wild type HEK293T CTx-647 stained and unstained cells as a control.

7. Gating is done to flow data based on size/shape of each event and viability based on DAPI staining (*see* **Notes 24** and **25**).

4 Notes

1. Small quantities of DNA are sent from Genecopia. It is best to transform competent cells to get higher quantities of DNA for transfection of target cell line.

2. The mCherry tag on the Genecopia plasmid is very dim in HEK293T cells, so it is best to use a small amount (1/10) EGFP DNA to track transfection efficiency.

3. CRISPR-Cas9 is dose dependent, but so are its potential off target effects. At lower efficiencies, such as the 1% efficiency we observed, off target effects are rarely seen.

4. OptiMem with Lipofectamine 2000 plus DNA can be changed anytime between 4–18 h with HEK293T cells. Little transfection related cell death occurs.

5. Lectins can target different glycoconjugates on the cell surface. GNL-Alexa 647 is chosen to make COG knockouts because it has been shown to give a more robust signal when COG function is impaired [23, 25]. GNL binds high-mannose residues, which would normally be at low levels on the cell surface, since these residues are further processed in the medial and trans Golgi.

6. Glycosylation defects can be seen beginning at 2 days, but we wait until transient expression of Cas9 (as assessed by the fluorophore mCherry or co-transfected EGFP) has diminished to assess glycosylation defects.

7. A confluency of ~50–80% is also the best to analyze the plasma membrane glycoproteins and lipids when doing lectin staining.

8. When plating cells on coverslips for lectin staining and immunofluorescence we have found that cells best adhere using collagen coating. This is also the best surface to plate the cells on for live-cell imaging.

9. Though cells plated on collagen adhere better, HEK293T cells are only semi-adherent so caution needs to be taken when staining or changing media. Additionally it is best for cells to be ~50–80% confluent when analyzing so they adhere better and do not come off in a sheet.

10. Some cell lines do not tolerate single cell sorting well. For this reason you may want to sort additional cells into a 3–6 cm dish to have a population enriched for glycosylation defects as a backup.

11. Cell sorting appears to be a step that can allow for contamination on occasion. For this reason we recommend sorting cells into wells containing media with 1% antibiotic/antimycotics.

12. Cells generally grow on the edge of wells, for this reason it is difficult to see colonies before 10 days.

13. Because HEK293T cells are semi-adherent we passage the cells from 96-well to 24-well for analysis simply by pipetting up and down to resuspend. We used a 1000 μL tip for this instead of a 200 μL pipet tip to cause less shearing forces on the cells.

14. Because chromosomal DNA does not require many cells, and it and whole cell lysates are both needed for validation, we usually do these steps in tandem.

15. When creating cell lysates for western blot analysis, the lysates are often quite viscous due to chromosomal DNA. To prevent this you can use a higher volume of 2%SDS, sonicate the lysate briefly, or boil the lysate for an additional minute or 2.

16. When preparing chromosomal DNA from the cells for PCR and sequencing, be sure to vortex well between steps.

17. For PCR we have found that 3–4 μL of chromosomal DNA and a few extra cycles in the thermocycler gives the best product yields.

18. After PCR we have found that gel purified PCR produce gives better sequencing results that just running the product through a PCR clean up kit.

19. As an additional verification step and to ensure that defects in KOs are from target specific mutations, we transiently transfect functioning COG protein back into the cells to ensure that it can correct any defects found.

20. SubAB is a potent toxin, so caution should be used while performing this assay. Dispose of toxin containing media in a sealed tube and place in a biohazard bag.

21. An additional PBS wash maybe desired to remove BSA contaminates before lysing.

22. We use the B subunit of cholera toxin, since this is a binding assay only. This enables the toxin to keep its lectin abilities, but is safer to work with.

23. We use DAPI as a viability dye because live cells take up the dye much slower than dead cells. Cells brightly stained for DAPI can be excluded from flow data by gating. Many other dyes, including Propidium Iodide, are also available for this purpose.

24. We try to make sure that there are at least 10,000 events after gating to analyze for CTx binding. For this reason it is helpful to gate when collecting data to be sure enough events have been obtained.

25. Here we used the NovoExpress software from ACEA Biosciences. FlowJo or other flow analysis software can also be used.

Acknowledgement

We would like to thank the Digital Microscopy, Flow Cytometry, and DNA Sequencing Core Facilities at UAMS for their help in this project. This work was supported, in part, by the NIH grants GM083144 and U54 GM105814.

References

1. Ungar D, Oka T, Brittle EE, Vasile E, Lupashin VV, Chatterton JE et al (2002) Characterization of a mammalian Golgi-localized protein complex, COG, that is required for normal Golgi morphology and function. J Cell Biol 157:405–415

2. Shestakova A, Zolov S, Lupashin V (2006) COG complex-mediated recycling of Golgi glycosyltransferases is essential for normal protein glycosylation. Traffic 7:191–204

3. Willett R, Kudlyk T, Pokrovskaya I, Schonherr R, Ungar D, Duden R et al (2013) COG complexes form spatial landmarks for distinct SNARE complexes. Nat Commun 4:1553

4. Ungar D, Oka T, Vasile E, Krieger M, Hughson FM (2005) Subunit architecture of the conserved oligomeric Golgi complex. J Biol Chem 280:32729–32735

5. Fotso P, Koryakina Y, Pavliv O, Tsiomenko AB, Lupashin VV (2005) Cog1p plays a central role in the organization of the yeast conserved oligomeric Golgi complex. J Biol Chem 280:27613–27623

6. Willett R, Ungar D, Lupashin V (2013) The Golgi puppet master: COG complex at center stage of membrane trafficking interactions. Histochem Cell Biol 140:271–283

7. Whyte JRC, Munro S (2001) The SeC34/35 Golgi transport complex is related to the exocyst, defining a family of complexes involved in multiple steps of membrane traffic. Dev Cell 1:527–537

8. Suvorova ES, Duden R, Lupashin VV (2002) The Sec34/Sec35p complex, a Ypt1p effector required for retrograde intra-Golgi trafficking, interacts with Golgi SNAREs and COPI vesicle coat proteins. J Cell Biol 157:631–643

9. Suvorova ES, Kurten RC, Lupashin VV (2001) Identification of a human orthologue of Sec34p as a component of the cis-Golgi vesicle tethering machinery. J Biol Chem 276:22810–22818

10. Kubota Y, Sano M, Goda S, Suzuki N, Nishiwaki K (2006) The conserved oligomeric Golgi complex acts in organ morphogenesis via glycosylation of an ADAM protease in C. elegans. Development 133:263–273

11. Ishikawa T, Machida C, Yoshioka Y, Ueda T, Nakano A, Machida Y (2008) EMBRYO YELLOW gene, encoding a subunit of the conserved oligomeric Golgi complex, is required for appropriate cell expansion and meristem organization in Arabidopsis thaliana. Genes Cells 13:521–535

12. Foulquier F (2009) COG defects, birth and rise! Biochim Biophys Acta 1792:896–902

13. Zeevaert R, Foulquier F, Jaeken J, Matthijs G (2008) Deficiencies in subunits of the Conserved Oligomeric Golgi (COG) complex define a novel group of Congenital Disorders of Glycosylation. Mol Genet Metab 93:15–21

14. Wu X, Steet RA, Bohorov O, Bakker J, Newell J, Krieger M et al (2004) Mutation of the COG complex subunit gene COG7 causes a lethal congenital disorder. Nat Med 10:518–523

15. Pokrovskaya ID, Szwedo JW, Goodwin A, Lupashina TV, Nagarajan UM, Lupashin VV (2012) Chlamydia trachomatis hijacks intra-Golgi COG complex-dependent vesicle trafficking pathway. Cell Microbiol 14:656–668

16. Liu S, Dominska-Ngowe M, Dykxhoorn DM (2014) Target silencing of components of the conserved oligomeric Golgi complex impairs HIV-1 replication. Virus Res 192:92–102

17. Zhu J, Davoli T, Perriera JM, Chin CR, Gaiha GD, John SP et al (2014) Comprehensive identification of host modulators of HIV-1 replication using multiple orthologous RNAi reagents. Cell Rep 9:752–766

18. Zolov SN, Lupashin VV (2005) Cog3p depletion blocks vesicle-mediated Golgi retrograde trafficking in HeLa cells. J Cell Biol 168:747–759

19. Kudlyk T, Willett R, Pokrovskaya ID, Lupashin V (2013) COG6 interacts with a subset of the Golgi SNAREs and is important for the Golgi complex integrity. Traffic 14:194–204

20. Laufman O, Freeze HH, Hong W, Lev S (2013) Deficiency of the Cog8 subunit in normal and CDG-derived cells impairs the assembly of the COG and Golgi SNARE complexes. Traffic 14:1065–1077

21. Jinek M, Chylinski K, Fonfara I, Hauer M, Doudna JA, Charpentier E (2012) A programmable dual-RNA-guided DNA endonuclease in adaptive bacterial immunity. Science 337:816–821

22. Smith RD, Willett R, Kudlyk T, Pokrovskaya I, Paton AW, Paton JC et al (2009) The COG complex, Rab6 and COPI define a novel Golgi retrograde trafficking pathway that is exploited by SubAB toxin. Traffic 10:1502–1517

23. Pokrovskaya ID, Willett R, Smith RD, Morelle W, Kudlyk T, Lupashin VV (2011) Conserved oligomeric Golgi complex specifically regulates the maintenance of Golgi glycosylation machinery. Glycobiology 21:1554–1569

24. Willett RA, Pokrovskaya ID, Lupashin VV (2013) Fluorescent microscopy as a tool to elucidate dysfunction and mislocalization of Golgi glycosyltransferases in COG complex depleted mammalian cells. Methods Mol Biol 1022:61–72

25. Ha JY, Pokrovskaya ID, Climer LK, Shimamura GR, Kudlyk T, Jeffrey PD et al (2014) Cog5-Cog7 crystal structure reveals interactions essential for the function of a multisubunit tethering complex. Proc Natl Acad Sci U S A 111:15762–15767

Chapter 13

Reversible Controlled Aggregation of Golgi Resident Enzymes to Assess Their Transport/Dynamics Along the Secretory Pathway

Riccardo Rizzo and Alberto Luini

Abstract

Golgi resident enzymes (GREs) are type II membrane proteins that are responsible for the processing of lipids and proteins within the Golgi stack, mostly by catalyzing the formation of long chains of different sugars bound to their substrates at specific positions. After synthesis and folding, GREs leave the ER to reach the Golgi complex where they are distributed along the Golgi stack in a fashion that reflects the sequential order of the sugar addition reactions they execute. Remarkably, the position of GREs within the stack is not stable; rather, the GREs appear to move rapidly across Golgi cisternae, perhaps to maintain the correct reaction order during the flux of traffic. It would thus be important to understand their dynamics, but methods to analyze their mobility within the stack are lacking. Here, we describe a tool to induce the reversible and controlled aggregation of GREs that can be used to study the ER-to-Golgi transport and intra Golgi dynamics of these enzymes (Rizzo et al., J Cell Biol 201:1027–1036, 2013; Tewari et al., Mol Biol Cell 26:4427–4437, 2015).

Key words Golgi, Glycosylation enzymes, Dynamics, Intra-Golgi transport, Regulated polymerization

1 Introduction

There are more than 250 mammalian GREs that all share a common general structure, which comprises a short cytosolic tail, a single transmembrane domain followed by a stem domain and a luminal globular catalytic domain [3]. The main function of these enzymes is to modify substrates by transferring and/or removing sugar moieties. This process, referred to as Golgi glycosylation, results in the attachment and elongation of long sugar chains on proteins and lipids.

While the dynamics and the mechanisms of transport of both transmembrane and soluble cargo proteins along the secretory pathway are well studied, much less is known about the same processes for GREs and especially about their intra-Golgi transport [4, 5], despite the fact that several molecules have been shown to

William J. Brown (ed.), *The Golgi Complex: Methods and Protocols*, Methods in Molecular Biology, vol. 1496,
DOI 10.1007/978-1-4939-6463-5_13, © Springer Science+Business Media New York 2016

contribute towards the proper Golgi localization of GREs [6]. This is unfortunate, because understanding the movement of the GREs along the secretory pathway and the contribution of the molecular players to this dynamics, is important to elucidate the mechanism of intra-Golgi transport and the regulation of lipid/protein glycosylation.

Here, we present a strategy and to monitor the intra-Golgi dynamics as well as the transport from ER to Golgi of the GREs. The strategy is based on use of constructs of GREs engineered to contain several copies (3–4) of a mutant form of FKBP referred to as FM domains [7] fused in tandem with the enzyme [1, 2]. FM domains spontaneously bind tightly to each other, causing polymerization of GREs and the formation of GRE aggregates. This type of aggregation appears to prevent the interaction of GREs, or of any other proteins, with the transport machinery [1, 8]. For instance, aggregation of a GRE in the ER prevents export of the enzyme from the ER to the Golgi and causes the accumulation of the polymerized enzyme in the ER [1]. This FM dependent aggregation can then be reversed by the addition of a synthetic drug that competes for the interacting surfaces of the FM domains and hence quickly causes the dissociation of the aggregates into single monomers (*see* also [9] for variations of this method). The synchronous disaggregation of the GRE construct exposes the GRE to the ER export machinery, promoting their synchronous exit from the ER and their final Golgi localization.

In a similar way, the aggregation of an engineered GRE construct within the Golgi will inhibit the interaction of the enzyme with the Golgi retention/recycling machinery causing the construct to follow passively the movements of the container in which it is localized, the Golgi cisternae. When the drug is later added to induce disaggregation at a time of interest, the enzyme construct will be synchronously exposed to the intra-Golgi transport machinery, allowing the experimenter to monitor the effects of such machinery on GRE transport.

In sum, through the manipulation of the aggregation state of the engineered GREs along the secretory pathway (in the ER as well as within the Golgi), the experimenter can synchronize exposure of the enzyme construct to the transport machinery and then monitor the dynamics and the localization of the transfected construct by microscopy. Monitoring transport within the stack requires the use of electron microscopy or of high resolution light microscopy.

We have applied this strategy [1] to two GREs: the Mouse α-1,2-mannosidase IB (MANI; [10]), a cis-medial (early) Golgi enzyme, and the trans (late) Golgi enzyme β-1,4-galactosyltransferase (GALT, [11]). The cytosolic, transmembrane and the luminal stem domain (Golgi-targeting portion) of MANI and GALT were cloned and three tandem-FM domains followed by an HA tag (hereafter referred to as MANI-FM and GALT-FM respectively) were fused to their C-terminal portion, and then used as described below.

Application of the strategy requires setting up proper conditions for inducing aggregation as well as methods to verify the aggregation state of the construct. The main steps are as follows

1. A biochemical assay to monitor the aggregation/disaggregation kinetics of the engineered GREs.

2. An imaging assay to monitor the aggregation state of the artificial engineered GREs in the Golgi.

3. An imaging assay to monitor the ER to Golgi dynamic of GREs.

2 Materials

2.1 Cell Culture and Transfection Reagents

1. Purified plasmid constructs encoding GALT-FM and MANI-FM are dissolved in TE buffer (10 mM Tris–HCl, 1 mM EDTA, pH 7.5) at a final concentration of 1 mg/mL. Store at –20 °C.

2. TransIT-LT1 transfection reagent (Mirus, USA). Store at 4 °C.

3. OptiMEM culture medium. Store at 4 °C.

4. RPMI complete medium: RPMI supplemented with 4.5 g/L glucose 2 mM glutamine, 100 U/ml penicillin, 100 µg/mL streptomycin, and 10 % fetal calf serum (FCS). Store at 4 °C.

2.2 Immunofluorescence Reagents

1. AP: synthetic drug named AP21998 (ARIAD Pharmaceuticals), currently available from Clontech (Takara Bio Inc). Store at –20 °C (see Note 1).

2. DPBS (10×), Dulbecco's Phosphate Buffered Saline: Prepare 1× solution with ultrapure water and store at 4 °C.

3. 24-well plates.

4. Glass coverslips.

5. Mounting media (16 % [w/v] Mowiol 4–24 in 30 % [v/v] glycerol in PBS).

6. Aqueous solution of 8 % paraformaldehyde in water.

7. Blocking solution: containing 50 mM NH$_4$Cl, 0.05 % saponin, 0.5 % BSA in PBS 1×, pH 7.4.

8. Standard microscope slides.

9. List of antibodies: Anti-HA monoclonal antibody (1:1000 for IF and 1:5000 for Western blotting) from Covance; anti-Mannosidase I (1:100 for IF); Alexa Fluor-conjugated secondary antibodies raised in donkey (Alexa Fluor 488, 568; 1:400 for IF) from Invitrogen; secondary antibodies conjugated with horseradish peroxidase (HRP) and directed against mouse (1:10,000) from Calbiochem (CA, USA).

10. Cycloheximide: 50 mM in absolute ethanol . Store at –20 °C.

11. Brefeldin A: dissolve in DMSO to have 10 µg/µL as final concentration. Store at –20 °C.

2.3 Reagents for Biochemistry

1. AP drug and DPBS reported in Subheading 2.2 (**items 1** and **2**).
2. 6-well plates.
3. Lysis buffer: 1% [v/v] Triton X-100, 20 mM Hepes, pH 7.4, 100 mM KCl, 2 mM EDTA, 1 mM dithiothreitol. Add protease inhibitor cocktail (Roche) only before the use. Store at 4 °C.
4. Cell scraper.

3 Methods

3.1 Biochemical Assay to Monitor in Time the Aggregation/Disaggregation Kinetics of the Engineered GREs

Cells can be transfected with one of the above FM-containing constructs and treated to induce aggregation and disaggregation [1]. The kinetic of aggregation/disaggregation of GALT-FM and MANI-FM can be followed by using a biochemical sedimentation assay, based on the principle that aggregates will precipitate with the pellet fraction after low speed centrifugation while monomer will remain in the supernatant fraction [1, 12].

1. Seed 200,000 HeLa cells onto 3.5-cm petri dishes (6 well plate) in RPMI complete medium (5% CO_2 at 37 °C).
2. After 24 h, prepare the transfection mix as follows: for each well add 100 µL of OptiMEM at 37 °C in an eppendorf tube; add 7.5 µL of TransIT-LT1, mix nicely with the pipette and wait for 3 min. Then, add 2.5 µg of the engineered GREs-FM cDNA (GALT-FM or MANI-FM) in the transfection mixture, shake gently and keep at room temperature (RT) for 15 min.
3. Change the media in each well with fresh RPMI complete medium (at 37 °C) in the presence or absence of 1 µM AP (based on how the experiment has been designed) to alter the polymerization status of the GREs-FM.
4. Leave the cells for 24 h in the incubator (5% CO_2 at 37 °C).
5. After 24 h, replace the media with a fresh media with or without AP and leave it for additional 30 min.
6. Place the cells on ice.
7. Remove the media and wash the cells 3–4 times with 2 mL of cold 1× PBS for each well.
8. Scrape the cells off using a cell scrapper in 1 mL/dish of DPBS 1×, transfer the cell suspension to a 1.5 mL eppendorf tube and centrifuge at $150 \times g$ for 5 min.
9. Remove the supernatant and resuspend the pellet in 100 µL of lysis buffer, and incubate the samples on ice for 20 min.
10. Centrifuge the lysate at $20,000 \times g$ for 10 min, and carefully separate the supernatant from the pellet fraction.
11. Add 100 µL of lysis buffer on the pellet fraction and resuspend the pellet.

12. Add sample buffer to both cell pellet extract and supernatant fractions, boil the samples for 5 min at 95 °C before loading and analyze by SDS-PAGE and immunoblotting.

See Fig. 1, for typical results expected from controlled polymerization/depolymerization of FM tagged constructs of GRE. To evaluate the result, the intensity of the bands are quantitated using Image J. The intensity of each band is reported as a percentage of the total, (for each experimental condition, the total is the sum of the intensities from pellet and supernatant fractions). As can be seen from Fig. 1, aggregation and disaggregation of Golgi enzymes occur within a few minutes [1].

3.2   An Imaging Assay to Monitor the Aggregation State of the Artificial Engineered GREs in the Golgi

Brefeldin A (BFA) is a fungal metabolite that blocks transport along the secretory pathway by causing the collapse of the Golgi into the ER [13]. When BFA is added to cells, it initially induces the formation of tubular intermediates that have a diameter (40–80 nm) similar to that of COPI-coated vesicles. These tubules, which contain GREs, fuse with the ER membrane and cause the redistribution of the Golgi residents in to the ER. The use of the BFA assay as a test to check the polymerization state of GREs-FM within the Golgi is based on the fact that GREs-FM polymer cannot physically enter into BFA-induced tubules [1] probably because GREs polymers are too large or form inflexible scaffolds that do not fit the curvature of the membrane tubules.

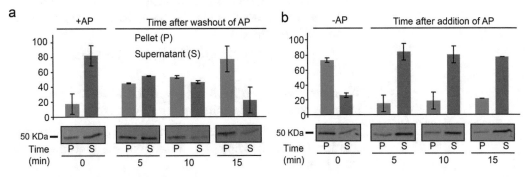

Fig. 1 A biochemical assay to monitor the aggregation/disaggregation kinetics of the engineered GREs. A biochemical assay to monitor the aggregation/disaggregation kinetics of the engineered GREs. (**a**) MANI-FM aggregation. HeLa cells were transiently transfected with MANI-FM in the presence of AP. After 24 h AP was washed out to induce the aggregation of artificial GRE and samples were lysed and centrifuged at low speed to separate the pellet from the supernatant fraction (*see* Subheading 3.1 for details). (**b**) MANI-FM disaggregation. HeLa cells were treated as in (**a**), and then after 24 h AP was washed out for 15 min. The drug was then added back (AP 1 μM) to cells before lysing at indicated time point. Samples were than processed for SDS-PAGE and Western Blotting. Antibody against HA was used to visualized MANI-FM. The aggregation (**a**) as well as disaggregation (**b**) kinetics is a fast process. Data are mean ± SEM. ©Rizzo et al., 2013. Originally published in Journal of Cell Biology. vol. 201 no. 7: 1027–1036

1. Seed 50,000 HeLa cells onto glass coverslips in 24-well plates in RPMI complete medium (5% CO_2 at 37 °C).

2. After 24 h, prepare the transfection mix as follows: for each well add 20 µL of OptiMEM at 37 °C in an eppendorf tube; add 1.5 µL of TransIT-LT1 and mix nicely with the pipette; wait for 3 min. Then, add 0.5 µg of the engineered GREs-FM cDNA (GALT-FM or MANI-FM) together with 0.5 µg of KDELR-GFP cDNA in the transfection mixture, shake gently and kept at RT for 15 min before adding onto cells.

3. Change the media with fresh RPMI complete media (at 37 °C) containing 1 µM AP.

4. Add the transfection mixture onto the cells.

5. After 24 h of transfection, change the media with a fresh one containing 1 µM AP and 50 µg/mL of cycloheximide at 37 °C, and leave it for 30 min (*see* **Note 2**).

6. Washout AP in the subset of samples where MAN-FM or GALT-FM have to aggregate, by doing three washes with PBS1× at 37 °C.

7. Replace with a fresh RPMI complete (?) containing 50 µg/mL of cycloheximide at 37 °C, only in the subset of samples subjected to aggregation.

8. Wait for 10 min to induce MAN-FM or GALT-FM aggregation in the Golgi (*see* **Note 3**).

9. Change the media with fresh RPMI complete medium at 37 °C, containing 50 µg/mL of cycloheximide, with or without AP depending on the samples, and containing 6 µg/mL of BFA, and leave it on cells for 6 or 8 min to induce Golgi tubulation [13] (*see* **Note 4**).

10. Fix the cells with 4% paraformaldehyde in DPBS1× for 10 min at RT.

11. Wash the glass coverslips 3 times with DPBS1× and permeabilize cells with blocking solution for 10 min at RT.

12. Incubate the cells with the anti-HA antibody at appropriate dilution (*see* Subheading 2.2, **item 9**) in the blocking buffer for 1 h at RT.

13. Wash the glass coverslips with DPBS1X and incubate with secondary antibodies labeled with appropriate fluorescence dyes (*see* Subheading 2.2, **item 9**), diluted in blocking solution, and leave for 1 h at RT.

14. Wash the coverslips 3 times with DPBS1× and finally with ultrapure water to remove PBS salt (which can affect imaging).

15. Mount the coverslips on glass microscope slides using the mounting media.

16. Leave at RT for 24 h in order to dry the sample before the confocal microscope analysis.

Experiments show that when GREs polymerize within the Golgi, their entry into BFA-induces tubules is inhibited (*see* Fig. 2; [1]). Thus, counting the numbers of Golgi tubules positive for GREs-FM, will be an imaging method to evaluate quantitatively the aggregation state of the artificial engineered GREs in the Golgi.

3.3   An Imaging Assay to Monitor the ER to Golgi Dynamic of GREs

The transient transfection of GREs-FM (MANI-FM or GALT-FM) in the absence of AP, causes the accumulation of these constructs in the ER [1]. The addition of AP onto cells, will cause fast depolymerization of the artificial GRE, which leave the ER and reach the Golgi (Fig. 3a).

1. Seed 50,000 HeLa cells onto glass coverslips in 24-well plates in RPMI complete medium (5 % CO_2 at 37 °C).

2. After 24 h, prepare the transfection mix as follows (for one well of a 24-well plate) add 20 μl of OptiMEM at 37 °C in an eppen-

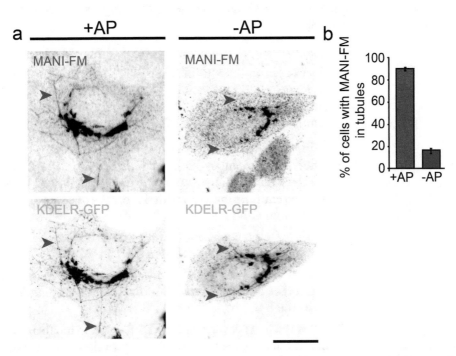

Fig. 2 An imaging assay to monitor the aggregation state of the artificial engineered GREs in the Golgi. An imaging assay to monitor the aggregation state of the artificial engineered GREs in the Golgi. (**a**) HeLa cells were transiently co-transfected with MANI-FM and KDEL receptor-GFP (KDELR-GFP) in the presence of AP. After 24 h, cells with and without AP were treated with BFA 6 μg/mL and fixed after 8 min. Samples were processed for immuno fluorescence (IF) and labeled with anti-HA antibody to visualize MANI-FM transfected cells (*see* Subheading 3.2 for details). (**b**) Quantifications were done by counting the number of KDELR-GFP containing tubules that are positive for MANI-FM. The formation of GRE aggregates within the Golgi, affects their entry into BFA-induced tubules. Data are mean ± SEM. Bars: 15 μm (**a**, +AP); 18 μm (**a**, −AP). ©Rizzo et al., 2013. Originally published in Journal of Cell Biology. vol. 201 no. 7 1027–1036

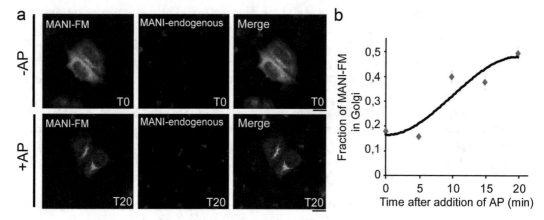

Fig. 3 An imaging assay to monitor the ER to Golgi dynamic of GREs. An imaging assay to monitor the ER to Golgi dynamic of GREs. HeLa cells were transiently transfected with MANI-FM in the absence of AP. After 24 h AP was added and samples were fixed at different time point. Samples were processed for IF and labeled with anti-HA and anti MANI-endogenous antibodies respectively (*see* Subheading 3.3 for details). (**b**) Exiting kinetics of MANI-FM from ER to the Golgi measured as integrated intensity of MANI-FM in the Golgi normalized to that of the ER. Values are mean ± SEM. Bars: 18 μm (**a**). ©Rizzo et al., 2013. Originally published in Journal of Cell Biology. vol. 201 no. 7 1027–1036

    dorf tube; add 1.5 μl of TransIT-LT1 and mix by pipetting; incubate for 3 min at RT. Then, add 0.5 μg of the plasmid (GALT-FM or MANI-FM) in the transfection mixture, shake gently and keep at RT for 15 min before adding it onto cells.

3. Leave the cells for 24 h in the incubator (5 % CO_2 at 37 °C).

4. After 24 h change the media with fresh RPMI complete medium (at 37 °C) containing 1 μM AP, followed by fixation at different time intervals with 4 % paraformaldehyde in PBS1× for 10 min at RT.

5. Wash the glass coverslips 3 times with DPBS1× and permeabilize cells with blocking solution for 10 min at RT.

6. Incubate the cells with the anti-HA and anti-Mannosidase I antibodies at appropriate dilutions (*see* Subheading 2.2, **item 9**) in the blocking buffer for 1 h at RT.

7. Wash the glass coverslips with DPBS1X and incubate with secondary antibodies labeled with appropriate fluorescence dyes (*see* Subheading 2.2, **item 9**), diluted in blocking solution, and leave for 1 h at RT.

8. Wash the coverslips 3 times with DPBS1× and finally with ultrapure water.

9. Mount the coverslips on glass microscope slides using the mounting media.

10. Leave at RT for 24 h in order to dry the sample before the confocal microscope analysis.

The measurement of fluorescence intensity in the Golgi of GREs-FM constructs over time can be monitored to assess the rate of ER to Golgi movement of the GREs (Fig. 3b).

4   Notes

1. We noticed that the AP is very sensitive to temperature and even a temporary storage on ice compromises its activity. So, when handling the drug for experimental use, we prefer to keep the drug in a mini cooler box (−20 °C).

2. The pretreatment with cycloheximide and AP for 30 min before starting the experiment provides a very clear Golgi localization of the engineered GREs. This is because cycloheximide, by blocking the protein synthesis, will clean the ER of newly synthesized proteins and the additional AP will ensure the complete depolymerization of the remaining aggregates present in the ER.

3. The $t1/2$ of GRE aggregation after AP washout is 10 min (as judged by biochemical assay, *see* Fig. 1).

4. The time required for BFA-induced Golgi tubulation depends on cell type and confluence. So it needs to be standardized before the planning of this experiment.

Acknowledgements

We thank Seetharaman Parashuraman for valuable discussions. We acknowledge the financial support of Italian Cystic Fibrosis Research Foundation (FFC#2 2014), Fondazione Telethon, AIRC (Italian Association for Cancer Research, IG 10593), the MIUR Project'FaReBio di Qualitá', the PON projects no. 01/00117 and 01-00862, PONa3-00025 (BIOforIU), PNR-CNR Aging Program 2012–2014 and Progetto Bandiera'Epigen' to AL. We thank the Bioimaging Facility at the Institute of Protein Biochemistry, Naples for help with image acquisition.

References

1. Rizzo R, Parashuraman S, Mirabelli P, Puri C, Lucocq J, Luini A (2013) The dynamics of engineered resident proteins in the mammalian Golgi complex relies on cisternal maturation. J Cell Biol 201:1027–1036

2. Tewari R, Bachert C, Linstedt AD (2015) Induced oligomerization targets Golgi proteins for degradation in lysosomes. Mol Biol Cell 26:4427–4437

3. Stanley P (2011) Golgi glycosylation. Cold Spring Harb Perspect Biol 3:1–14

4. Harris SL, Waters MG (1996) Localization of a yeast early Golgi mannosyltransferase, Och1p, involves retrograde transport. J Cell Biol 132:985–998

5. Opat AS, Houghton F, Gleeson PA (2001) Steady-state localization of a medial-Golgi glycosyltransferase involves transit through the trans-Golgi network. Biochem J 358:33–40

6. Banfield DK (2011) Mechanisms of protein retention in the Golgi. Cold Spring Harb Perspect Biol 3:a005264

7. Rivera VM, Wang X, Wardwell S, Courage NL, Volchuk A, Keenan T, Holt DA, Gilman M, Orci L, Cerasoli F Jr, Rothman JE, Clackson T (2000) Regulation of protein secretion through controlled aggregation in the endoplasmic reticulum. Science 287:826–830

8. Lavieu G, Zheng H, Rothman JE (2013) Stapled Golgi cisternae remain in place as cargo passes through the stack. Elife 2, e00558

9. Wong M, Munro S (2014) Membrane trafficking. The specificity of vesicle traffic to the Golgi is encoded in the golgin coiled-coil proteins. Science 346:1256898

10. Becker B, Haggarty A, Romero PA, Poon T, Herscovics A (2000) The transmembrane domain of murine alpha-mannosidase IB is a major determinant of Golgi localization. Eur J Cell Biol 79:986–992

11. Cole NB, Smith CL, Sciaky N, Terasaki M, Edidin M, Lippincott-Schwartz J (1996) Diffusional mobility of Golgi proteins in membranes of living cells. Science 273:797–801

12. Volchuk A, Amherdt M, Ravazzola M, Brugger B, Rivera VM, Clackson T, Perrelet A, Sollner TH, Rothman JE, Orci L (2000) Megavesicles implicated in the rapid transport of intracisternal aggregates across the Golgi stack. Cell 102:335–348

13. Lippincott-Schwartz J, Donaldson JG, Schweizer A, Berger EG, Hauri HP, Yuan LC, Klausner RD (1990) Microtubule-dependent retrograde transport of proteins into the ER in the presence of brefeldin A suggests an ER recycling pathway. Cell 60:821–836

Chapter 14

Assays to Study the Fragmentation of the Golgi Complex During the G2–M Transition of the Cell Cycle

Inmaculada Ayala and Antonino Colanzi

Abstract

The Golgi complex of mammalian cells is composed of stacks of flattened cisternae that are connected by tubules to form a continuous membrane system, also known as the Golgi ribbon. At the onset of mitosis, the Golgi ribbon is progressively fragmented into small tubular-vesicular clusters and it is reconstituted before completion of cytokinesis. The investigation of the mechanisms behind this reversible cycle of disassembly and reassembly has led to the identification of structural Golgi proteins and regulators. Moreover, these studies allowed to discover that disassembly of the ribbon is necessary for cell entry into mitosis. Here, we describe an in vitro assay that reproduces the mitotic Golgi fragmentation and that has been successfully employed to identify many important mechanisms and proteins involved in the mitotic Golgi reorganization.

Key words Golgi complex, Cell permeabilization, Mitosis, Immunofluorescence

1 Introduction

The Golgi complex is a central organelle of the secretory pathway, and is involved in the modification, sorting, and transport of proteins and lipids [1]. Furthermore, Golgi membranes can be also considered as a hub for the integration of various signaling pathways [1, 2].

The structural organization of this organelle varies greatly between species. For example, in the yeast *Saccharomyces cerevisiae* the Golgi complex consists of isolated cisternae that are distributed in the cytoplasm; the protozoan parasite *Toxoplasma gondii*, possesses an individual stack of cisternae; Drosophila and plants possess well-defined stacks that are dispersed throughout the cytosol [3, 4]. In contrast, in mammalian cells the Golgi complex is organized as an interconnected pericentriolar ribbon in interphase, which undergoes extensive fragmentation in the form of dispersed tubular-reticular and vesicular elements in mitosis [5]. This reversible disassembly can be schematically described as composed of three main steps. First, during the G2 phase of cell cycle the Golgi ribbon is disconnected to yield separate Golgi stacks (Fig. 1) [6]. Next, at the onset

William J. Brown (ed.), *The Golgi Complex: Methods and Protocols*, Methods in Molecular Biology, vol. 1496,
DOI 10.1007/978-1-4939-6463-5_14, © Springer Science+Business Media New York 2016

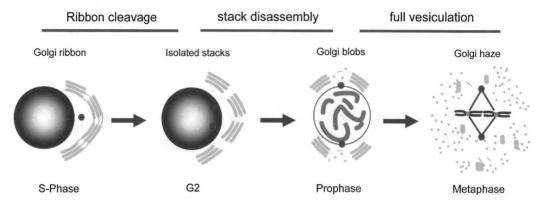

Fig. 1 The fragmentation of the Golgi ribbon during G2–mitosis. Schematic representation of the steps leading to breakdown of the Golgi complex in mitosis. The mammalian Golgi complex in S-phase forms a ribbon next to the centrosome (*red cylinders*) and the nucleus (*blue*). In G2, the interconnected ribbon is converted into isolated stacks. This step is necessary and sufficient for entry into mitosis (prophase). In prophase, the isolated Golgi stacks undergo further disassembly into the Golgi "blobs" and "haze"

of mitosis the stacks are converted into tubular-reticular membranes, termed "Golgi blobs" (Fig. 1), which are dispersed throughout the cytosol [7, 8]. Finally, the "Golgi blobs" are broken down by vesiculation to yield the "Golgi haze" (Fig. 1) [7]. The Golgi membranes are then reassembled in telophase and this involves the formation of two smaller Golgi ribbons that ultimately coalesce [9].

The investigations of the mechanism of mitotic Golgi fragmentation led to the discovery that cleavage of the Golgi ribbon is crucial for entry of cells into mitosis [10, 11], implying that cell cycle progression depends not only on the state of the DNA but also on the organization of an organelle. In this chapter, we describe in detail an in vitro assay that has been applied with success to identify many of the key proteins involved in this specific fragmentation step.

2   Materials

Prepare all solutions using ultrapure water and analytical grade reagents. Prepare and store all reagents at –20 °C (unless indicated otherwise). Diligently follow all waste disposal regulations when disposing waste materials.

2.1   General Reagents

1. Normal rat kidney (NRK) and HeLa cells and reagents/equipment for their maintenance. NRK and HeLa cells are grown in Dulbecco's modified Eagle's medium and minimal essential medium, respectively (Invitrogen), supplemented with 10 % fetal calf serum, 100 µM minimal essential medium nonessential amino acids solution, 2 mM L-glutamine, 1 U/ml penicillin, and 50 µg/ml streptomycin.

2. Mitotic extraction buffer (MEB): 15 mM HEPES (pH 7.4), 50 mM KCl, 10 mM MgCl$_2$, 20 mM β-mercaptoethanol, 20 mM β-glycerophosphate, 15 mM EGTA, 0.5 mM spermidine, 0.2 mM spermine, 1 mM DTT, 0.1 mM PMSF, 0.2 µg/ml aprotinin, 0.2 µg/ml leupeptin, and 0.2 µg/ml pepstatin.

3. KHM buffer: 25 mM HEPES–KOH (pH 7.2), 125 mM potassium acetate, 2.5 mM magnesium acetate.

4. KHM–KCl: KHM buffer with 1 M of KCl.

5. ATP-regenerating system: final concentration of the components: 0.5 mM ATP, 0.3 mM UTP, 10 mM creatine phosphate, and 12 U/ml creatine kinase. Store at –20 °C as 10× stock solution.

6. Fibronectin: we use fibronectin from human plasma at a final concentration of 10 µg/ml. Prepare a stock in PBS with 0.9 mM CaCl$_2$ and 0.5 mM MgCl$_2$. Store at –20 °C as 10× or 100× aliquots.

7. Stock of digitonin: prepare a stock solution of 10 mg/ml in DMSO. Use a final concentration of 30 µg/ml in ice-cold KHM buffer.

8. Stock of formaldehyde: prepare a solution of 4% formaldehyde in PBS. Store at 4 °C and protected from light.

9. Stock of blocking buffer for immunofluorescence: prepare a solution of 0.1% BSA, 0.1% saponin, 50 mM NH$_4$Cl, and 0.02% NaN$_3$ in PBS.

10. 2 mM thymidine. You can prepare stocks of 100 mM and store at –20 °C.

11. Nocodazole: stock solution of 33 µM in DMSO.

2.2 Cell Cycle Synchronization and Microinjection (Subheading 3.4.1)

1. Aphidicolin: stock solution of 20 mM in DMSO.

2. Fixable, fluorescently labeled dextran of 10,000 Da.

3. DAPI (stock of 5 mg/ml in water) or HOECHST (10 mg/ml in water).

2.3 Immuno-fluorescence Detection of Cells in S Phase (Subheading 3.4.2)

1. Bromodeoxyuridine (BrdU): stock solution of 10 mM in DMSO.

2. 4% paraformaldehyde and 0.5% TX-100 in PBS, pH 7.4.

3. 0.05% Tween 20 in PBS, pH 7.4.

4. 4 M of HCl.

5. 0.1 M sodium borate and 0.05% Tween 20 in PBS, pH 7.4.

6. Anti-BrdU antibodies.

2.4 Immuno-fluorescence Detection of Cells in G2 Phase

1. Antibodies against the phosphorylated form (Ser10) of Histone H3 (Millipore).

2. RO3306 (CDK inhibitor): stock solution of 18 mM in DMSO.

3 Methods

3.1 General Considerations About How to Monitor the Structure of the Golgi Ribbon

The key step of mitotic Golgi disassembly to allow G2–M transition is the G2-specific conversion of the interconnected Golgi ribbon into isolated stacks, which requires the cleavage of the membrane tubules connecting the stacks. This cleavage depends on at least three factors, which have all been characterized through the assay described below. They are the fission inducing protein CtBP1-S/BARS (referred to as BARS) [11] and the peripheral Golgi proteins GRASP65 and GRASP55 [10, 12]. Each of these proteins has a specialized role in Golgi ribbon breakdown and their combined activities produce isolated stacks in G2. GRASP55 is controlled by a kinase cascade composed of PKD [13] and RAF/MEK/ERK [14, 15] kinase cascade; while GRASP65 is controlled by JNK2 [16] and PLK1 [17].

It is important to note that these morphological changes appear minor and a cleaved but clustered group of Golgi ministacks can be indistinguishable at the fluorescence microscopy level from an intact ribbon. Thus, a cleaved ribbon can only be detected with sensitive approaches, such as Fluorescence recovery after photobleaching (FRAP) and electron microscopy [6].

3.2 Semi-intact Cell Assay to Study Mitotic Golgi Fragmentation

The experimental system that is described here is an assay that reconstitutes mitotic Golgi fragmentation in permeabilized cells. This assay has been developed in the laboratory of Dr. Vivek Malhotra, at the University of San Diego, California [18].

According to this assay, normal rat kidney (NRK) cells are permeabilized with digitonin, washed with high-salt-containing buffer to remove cytosolic proteins, then incubated with mitotic extracts prepared from NRK cells arrested in mitosis. After incubation at 32 °C, the organization of the Golgi membranes is monitored by fluorescence microscopy using Golgi-specific antibodies. In the presence of mitotic cytosol, the pericentriolar Golgi apparatus is converted into small fragments that are found dispersed in the cytoplasm [18]. The in vitro ultrastructure of these fragments has been characterized, demonstrating that the Golgi membranes are transformed into tubulo-reticular clusters that are similar to the Golgi fragments observed in intact NRK cells at prometaphase [19]. This assay allows a quantitative readout expressed as the percentage of cells with fragmented Golgi, and it has proven to be a very powerful tool because it reproduces mitotic Golgi fragmentation with high fidelity, while also allowing a great variety of manipulations of the experimental conditions to be made, such use of chemical inhibitors, addition of blocking antibodies, depletion of proteins of interest from the mitotic cytosol. A similar assay has been reproduced in MDCK cells [20].

3.2.1 Preparation of Mitotic Cytosol

The extracts are prepared according the method of Nakagawa [21].

1. Grow NRK cells in 15 cm petri dishes (about 70 % confluency). Prepare 40 dishes to obtain 1 ml of cytosol at about 12–14 mg/ml.

2. Induce an S-phase arrest with 2 mM thymidine (up to 16 h).

3. Remove thymidine with two washes with 15 ml of PBS and incubate with 500 ng/ml nocodazole in complete medium for up to 14 h to arrest cells in prometaphase (*see* **Note 1**). Typically, this procedure leads to the accumulation of more than 80 % of the cells in prometaphase. The mitotic cells will appear roundish and translucent under the light microscope.

4. Harvest round (mitotic) cells by "shake off" (*see* **Note 2**), collect the suspension in 50 ml falcon tubes and centrifuge at 500×*g* for 10 min.

5. Wash mitotic cells by resuspending the pellet with 10 ml of cold PBS and centrifuging the suspension at 500×*g* for 5 min at 4 °C. Repeat this operation twice.

6. Resuspend cell pellet in twice the cell bed volume using MEB in an Eppendorf tube. Centrifuge the pellet at 500×*g* for 5 min. Resuspend again the pellet in twice the cell bed volume using MEB buffer and incubate for 10 min on ice. As MEB buffer is hypotonic this will induce cell swelling.

7. Homogenize by repeated passages through an 18-gauge needle and through a 24-gauge needle (20–30 passages may be necessary). The purpose here is to breakup the plasma membrane without damaging the nuclei. Monitor homogenization state under the light microscope. The addition of trypan blue can help the visualization of broken cells (*see* **Note 3**).

8. Remove cell debris by high-speed centrifugation (100,000×*g* for 45 min at 4 °C).

9. Collect the supernatant (a typical preparation should have a protein concentration of about 12–14 mg/ml) and freeze in aliquots (about 25 μl) in liquid nitrogen.

10. Store at −80 °C (*see* **Note 4**).

3.2.2 Preparation of Interphase Cytosol

1. Grow NRK cells in 15 cm petri dishes (70 % confluency). Prepare 40 dishes.

2. Arrest cells in S-phase with 2 mM thymidine for up to 16 h.

3. Wash cells with cold PBS to remove thymidine.

4. Harvest the cells with gentle scraping by using a rubber scraper (*see* **Note 5**), collect the suspension in 50 ml falcon tubes and centrifuge at 500×*g* for 10 min.

5. Resuspend cell pellet in twice the cell bed volume using MEB in an Eppendorf tube. Centrifuge the pellet at 500×*g* for 5 min at 4 °C. Resuspend again the pellet in twice the cell bed

volume using MEB buffer and incubate for 10 min on ice. As MEB buffer is hypotonic this will induce cell swelling.

6. Carry out the **steps 7–10** described for the preparation of Mitotic Extract (section 3.2.1).

3.2.3 Selective Permeabilization of Plasma Membrane of NRK Cells

1. Coat coverslips with fibronectin (*see* **Note 6**).

2. Grow NRK cells on fibronectin-coated coverslips (60 % confluency) in a 24-well plate.

3. Incubate the cells with 2 mM thymidine for 6–16 h (*see* **Note 7**) before the experiment.

4. Wash cells with KHM buffer at room temperature.

5. Move the plate on ice and wash twice with cold KHM buffer (*see* **Note 8**).

6. Permeabilize the plasma membrane by treating the cells with 30 μg/ml digitonin in cold KHM for 3–5 min. The exact timing of digitonin treatment has to be empirically determined (*see* **Note 9**).

7. Stop permeabilization by washing the cells with ice-cold KHM.

8. Incubate permeabilized cells with ice-cold KHM-KCl for 5 min to remove peripheral membrane proteins.

9. Allow permeabilized cells to warm up during the last 2 min of the salt wash by placing the 24-well plate on the bench.

10. Wash the cells with KHM buffer at room temperature (*see* **Note 10**).

3.2.4 Incubation of Permeabilized Cells with Cytosol

1. Mix 25 μl of mitotic or interphase cytosol (12–14 mg/ml) with 5 μl of ATP-regeneration system 10×.

2. Bring each sample to a final volume of 50 μl using KHM buffer and the item to be tested (e.g., antibody, chemical inhibitors) dissolved in either PBS or DMSO. The presence of PBS (up to 5 μl) and of DMSO (up to 1 μl) in the reaction mixture does not affect the assay.

3. Place the reaction mixture as a single drop on Parafilm located on a 10 cm tissue culture dish (Fig. 2A) (*see* **Note 11**).

4. Place inverted coverslips with permeabilized cells onto the reaction mixture (Fig. 2B), cover the plate, and incubate in a water bath for 60 min at 32 °C.

5. Fix the cells by incubating with 4 % paraformaldehyde for 10 min (*see* **Note 12**).

6. Prepare the samples for immunofluorescence analysis (*see* **Note 12**).

7. Monitor Golgi fragmentation by counting the fraction of cells with fragmented Golgi (Fig. 3).

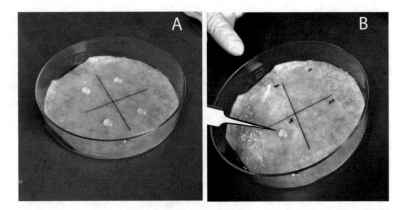

Fig. 2 The incubation of permeabilized cells with cytosol. Place the reaction mixture as a single drop on Parafilm located on a 10 cm tissue culture dish on a top of wet tissue (**A**). With the help of tweezers place the inverted coverslips with permeabilized cells onto the reaction mixture. (**B**). Cover and incubate on a water bath

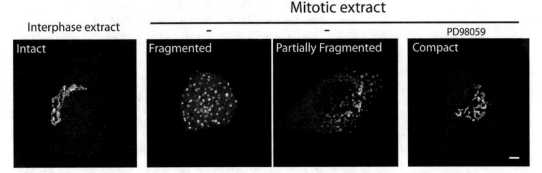

Fig. 3 Morphology of the Golgi complex in permeabilized cells. Digitonin-permeabilized NRK cells were incubated with interphase extract or mitotic extracts in the presence of DMSO (–) or of 80 μM of the MEK inhibitor PD98059. The cells were fixed and prepared for immunofluorescence analysis. An anti-Giantin antibody has been used to identify the Golgi membranes. Note that the cells incubated with mitotic extracts can show variable levels of Golgi fragmentation. Bar: 10 μm

3.3   Cell-Free Assays to Study Mitotic Golgi Fragmentation and Reassembly

Mitotic Golgi disassembly and post-mitotic reassembly have been reconstituted using purified Golgi membranes, and mitotic cytosol or purified recombinant proteins. The ultrastructure of Golgi membranes is examined by electron microscopy. These assays have been developed in the laboratory of Dr. G. Warren [22, 23] and we refer to excellent reviews describing these assays and detailed protocols [24, 25].

3.4 Additional Supporting Methods for the Study of Mitotic Golgi Fragmentation

3.4.1 Cell Cycle Synchronization of NRK/ HeLa Cells and Microinjection

It is important to support the results obtained through the assay with experiments performed in intact, living cells. The "canonical" treatments (e.g., siRNA treatment or protein overexpression) are effective and they can affect a large cell population, but in general may result in smaller and/or secondary effects. Thus, in our experience a successful protocol to block the cleavage of the Golgi ribbon is the microinjection of "blocking reagents" (e.g., antibodies, recombinant proteins and peptides), which target proteins involved in the process [10, 15]. This approach is essential to induce an acute block of Golgi fragmentation and, as a consequence, a potent and prolonged G2 arrest. However, the drawback of this approach is that it is limited to the observation of single cells by immunofluorescence.

1. Grow NRK or HeLa cells on fibronectin-coated glass coverslips (60% confluency).

2. Synchronize NRK cells at the G1/S transition by treating them with 2.5 μg/ml aphidicolin in complete medium for 16 h (*see* **Note 13**). HeLa cells are better synchronized by two consecutive overnight treatments with 2 mM thymidine.

3. Remove aphidicolin (or thymidine) by washing three times for 3 min with warm complete medium.

4. Two hours after removal of the S-phase block, inject 200 cells with purified recombinant proteins or antibodies (8–12 mg/ml) in the presence of fixable fluorescently labeled dextran as an injection marker.

5. Continue the incubation until the mitotic peak in control cells (7–8 h for NRK cells; 10–13 h for HeLa cells), fix and process for immunofluorescence as required and with DAPI or HOECHST to stain the DNA. To reduce the inter-coverslip variations, a "relative mitotic index" is calculated as percentage of microinjected cells in mitosis normalized to non-microinjected cells on the same coverslip [6].

3.4.2 Immuno-fluorescence-Based Detection of Cells in S Phase

A useful and simple approach for detecting cells in S-phase is the bromodeoxyuridine (BrdU) staining [10]. BrdU is an analog of thymidine that is incorporated into newly synthesized DNA during replication.

1. Incubate NRK or HeLa cells with 15 μM BrdU in maintenance culture medium for 30 min (*see* Subheading 2).

2. Fix and permeabilize cells in 4% paraformaldehyde and 0.5% Triton X-100 in PBS for 10 min.

3. Wash quickly cells with 0.05% Tween in PBS twice.

4. Denature DNA with 4 M of HCl for 30 min.

5. Wash cells with 0.1 M sodium borate and 0.05% Tween in PBS.

6. The cells then can be labeled with anti-BrdU antibodies using standard immunofluorescence methods (*see* ref. 10 for more details).

3.4.3 Immuno-fluorescence-Based Detection of Cells in the G2 Phase

The identification of NRK and HeLa cells in the G2 phase of cell cycle can be accomplished by labeling with an antibody against the phosphorylated form (Ser10) of Histone H3. The phosphorylation of histone H3 is first detected within pericentromeric heterochromatin during G2 and then spreads to the entire chromosome in coincidence with mitotic chromosome condensation. Thus, G2 cells can be identified for the presence of labeling at the level of centromeres and for the absence of chromosome condensation (monitored by DAPI or HOECHST staining) [6]. If needed, cells can be blocked at the G2 phase of the cell cycle by an overnight treatment with 9 μM of RO3306 (CDK inhibitor) or by incubating cells with 1 μg/ml of Hoechst 33342 or 40 μg/ml of Hoechst 33258 for 18 h [6, 16].

3.5 Concluding Remarks

This chapter describes in detail an in vitro assay that has been instrumental to the identification of many proteins involved in regulating Golgi disassembly during mitosis. The complex ribbon organization of the mammalian Golgi apparatus depends on a wide range of proteins, among which the Golgins, which are peripheral membrane proteins that are proposed to provide a structural skeleton [26]. The peripheral Golgi proteins GRASP65 and GRASP55 also contribute to Golgi ribbon formation [27, 28]. Golgi ribbon organization also depends on an intact microtubule and actin cytoskeleton, specialized cytoskeleton-based motors, and membrane input from the endoplasmic reticulum (ER) [29–31], but these elements are not discussed here in detail.

Thus, the complexity and multiplicity of proteins and signaling networks involved in maintaining the structural organization of the Golgi can complicate the interpretation of the results obtained by siRNA or transfection approaches, which can give rise to adaptation and compensatory mechanism of cells during long-term treatments. Therefore, we think that the assay described here offers a valuable experimental approach to study many of the mechanisms involved in regulating the structure of the Golgi ribbon. However, there is also a need for the development of methods to rapidly inactivate the function of proteins involved in Golgi organization in an entire cell population. Such approaches will help to understand the timing and temporal order in which structural Golgi components and regulatory kinases act in concert to assemble and disassemble the Golgi ribbon, and to employ proteomic and/or phosphoproteomic approaches to gain a more global view of the process. Perhaps an interesting experimental strategy toward this direction has been recently applied in the A. Linstedt's group, and consists of transfecting the protein of interest fused with a KillerRed tag, which allows the rapid inactivation of the protein of interest upon light irradiation (*see* ref. 32 and references herein).

4   Notes

1. Treatment with nocodazole for long periods of time can induce apoptosis. Be careful and monitor the cells for signs of apoptosis. In our experience NRK cells start to show evidence of apoptosis after 12–16 h of incubation.

2. To facilitate the harvest of round (mitotic) cells, remove the medium, place it in a 50 ml falcon tube and gently beat the border of the Petri dish on a side of the bench. We generally repeat the beatings up to 20–30 times or until more than 80 % of the rounded cells are detached. To collect the detached cells wash twice the petri dish with 5 ml of ice cold PBS and mix the suspension with the previously collected medium.

3. Take a few microliters of the suspension, mix with trypan blue, and put the drop on a glass slide and cover it with a coverslip. Effectively broken cells will appear as isolated nuclei surrounded by small debris.

4. Cytosol can be stored for up to 2 years at –80 °C without loosing activity. The cytosol can be used after freeze-thaw cycles (up to twice) without loss of activity.

5. Be very gentle in scraping. The goal here is to detach the cells without damaging them.

6. For fibronectin coating of coverslips, place a drop (about 20 μl) on the top of each coverslip, incubate it for 15 min at room temperature, wash with PBS, and let dry under the hood.

7. The use of thymidine reduces the number of cells in mitosis.

8. Place the 24-wells plate on a bucket full of "wet" ice to speed the cooling.

9. We have tested other permeabilization methods (such as Streptolysin O) without success. Digitonin is a non-ionic detergent that specifically permeabilizes membranes enriched in cholesterol (such as the plasma membrane). The only intracellular compartment rich in cholesterol is the trans-Golgi network. Please, keep in mind that the quality of digitonin can be very different from batch to batch. As a general characterization for finding the exact experimental conditions, we perform first an incubation of the permeabilized cells with an anti-antibody against a cytoplasmically oriented Golgi protein (e.g., GRASP65 or Giantin) prior to fixation. This procedure allows the selective staining of permeabilized cells.

10. As general advice, we suggest to perform experiments involving a maximum of six experimental points as it is difficult to manage more samples. It is better to perform a test of permeabilization before the actual experiment, because in our experience, the time can change a little bit from one day to another. When you permeabilize the cells, check coverslips that are

most similar (exclude coverslips with too many detached cells). Permeabilized cells will appear flat and translucent under the light microscope. Learn to evaluate them by using different objectives. Prepare more coverslips than needed to select comparable ones and discard the ones with aggregated cells.

11. The mitotic cytosol can be "manipulated" by adding a chemical inhibitor for a protein kinase that is involved in Golgi disassembly. We recommend to perform experiments to estimate its efficacy as the vast majority of inhibitors are competitive for the ATP binding site and the concentration of ATP in the assay is high (500 μM). The mitotic cytosol can also be treated with an antibody to deplete a protein of interest (*see* ref. 11) or subjected to chromatographical fractionation (*see* ref. 18). In our experience the cytosol does not lose its activity for up to 16 h if all the procedures are performed at 4 °C.

12. Be very gentle during the fixation step; we usually "lift" the coverslip by directly pipetting formaldehyde under the coverslip, picking them up, and placing them in a 24-well plate (cell side up). All the subsequent washes are made minimizing them and gently removing the buffer with a pipet. For example, we recommend avoid washes before the addition of blocking buffer and before the addition of the primary antibody. We then perform one wash in PBS before the addition of the secondary antibody and two washes in PBS before mounting the coverslips. Typically, after the permeabilization and incubation many cells will be detached. *See* Fig. 3 for an example of a fragmented and intact Golgi. Always add a positive control of inhibition (e.g., 80 μM of PD98059, an inhibitor of MEK) to induce a block of Golgi fragmentation. The typical fraction of cells with fragmented (including partially fragmented) Golgi varies from a 60 to 90 %.

13. Aphidicolin is an antibiotic isolated directly or as a secondary metabolite of the fungi *Cephalosporium aphidicola* and *Nigrospora oryzae*, respectively. It is a reversible inhibitor of DNA replication, thus blocking the cell cycle at the S phase of eukaryotic cells or some viruses by inhibiting the DNA polymerase II or the induced DNA polymerases, respectively. For NRK cells it is used at 2.5 μg/ml for 16 h, while in HeLa cells it can induce apoptosis [33].

Acknowledgments

We apologize to those colleagues whose work we were not able to discuss due to space limitation. A.C. acknowledges the Italian Association for Cancer Research (AIRC, Milan, Italy; IG6074) for financial support.

References

1. Wilson C, Venditti R, Rega LR, Colanzi A, D'Angelo G, De Matteis MA (2011) The Golgi apparatus: an organelle with multiple complex functions. Biochem J 433:1–9

2. Cancino J, Luini A (2013) Signaling circuits on the Golgi complex. Traffic 14:121–134

3. Preuss D, Mulholland J, Franzusoff A, Segev N, Botstein D (1992) Characterization of the Saccharomyces Golgi complex through the cell cycle by immunoelectron microscopy. Mol Biol Cell 3:789–803

4. Pelletier L, Stern CA, Pypaert M, Sheff D, Ngo HM, Roper N, He CY, Hu K, Toomre D, Coppens I, Roos DS, Joiner KA, Warren G (2002) Golgi biogenesis in Toxoplasma gondii. Nature 418:548–552

5. Misteli T, Warren G (1995) Mitotic disassembly of the Golgi apparatus in vivo. J Cell Sci 108:2715–2727

6. Colanzi A, Hidalgo Carcedo C, Persico A, Cericola C, Turacchio G, Bonazzi M, Luini A, Corda D (2007) The Golgi mitotic checkpoint is controlled by BARS-dependent fission of the Golgi ribbon into separate stacks in G2. EMBO J 26:2465–2476

7. Colanzi A, Suetterlin C, Malhotra V (2003) Cell-cycle-specific Golgi fragmentation: how and why? Curr Opin Cell Biol 15:462–467

8. Nelson WJ (2000) W(h)ither the Golgi during mitosis? J Cell Biol 149:243–248

9. Gaietta GM, Giepmans BN, Deerinck TJ, Smith WB, Ngan L, Llopis J, Adams SR, Tsien RY, Ellisman MH (2006) Golgi twins in late mitosis revealed by genetically encoded tags for live cell imaging and correlated electron microscopy. Proc Natl Acad Sci U S A 103:17777–17782

10. Sutterlin C, Hsu P, Mallabiabarrena A, Malhotra V (2002) Fragmentation and dispersal of the pericentriolar Golgi complex is required for entry into mitosis in mammalian cells. Cell 109:359–369

11. Hidalgo Carcedo C, Bonazzi M, Spano S, Turacchio G, Colanzi A, Luini A, Corda D (2004) Mitotic Golgi partitioning is driven by the membrane-fissioning protein CtBP3/BARS. Science 305:93–96

12. Xiang Y, Wang Y (2010) GRASP55 and GRASP65 play complementary and essential roles in Golgi cisternal stacking. J Cell Biol 188:237–251

13. Kienzle C, Eisler SA, Villeneuve J, Brummer T, Olayioye MA, Hausser A (2013) PKD controls mitotic Golgi complex fragmentation through a Raf-MEK1 pathway. Mol Biol Cell 24:222–233

14. Feinstein TN, Linstedt AD (2007) Mitogen-activated protein kinase kinase 1-dependent Golgi unlinking occurs in G2 phase and promotes the G2/M cell cycle transition. Mol Biol Cell 18:594–604

15. Persico A, Cervigni RI, Barretta ML, Corda D, Colanzi A (2010) Golgi partitioning controls mitotic entry through Aurora-A kinase. Mol Biol Cell 21:3708–3721

16. Cervigni RI, Bonavita R, Barretta ML, Spano D, Ayala I, Nakamura N, Corda D, Colanzi A (2015) JNK2 controls fragmentation of the Golgi complex and the G2/M transition through phosphorylation of GRASP65. J Cell Sci 128:2249–2260

17. Sengupta D, Linstedt AD (2010) Mitotic inhibition of GRASP65 organelle tethering involves Polo-like kinase 1 (PLK1) phosphorylation proximate to an internal PDZ ligand. J Biol Chem 285:39994–40003

18. Acharya U, Mallabiabarrena A, Acharya JK, Malhotra V (1998) Signaling via mitogen-activated protein kinase kinase (MEK1) is required for Golgi fragmentation during mitosis. Cell 92:183–192

19. Colanzi A, Deerinck TJ, Ellisman MH, Malhotra V (2000) A specific activation of the mitogen-activated protein kinase kinase 1 (MEK1) is required for Golgi fragmentation during mitosis. J Cell Biol 149:331–339

20. Kano F, Nagayama K, Murata M (2000) Reconstitution of the Golgi reassembly process in semi-intact MDCK cells. Biophys Chem 84:261–268

21. Nakagawa J, Kitten GT, Nigg EA (1989) A somatic cell-derived system for studying both early and late mitotic events in vitro. J Cell Sci 94:449–462

22. Rabouille C, Misteli T, Watson R, Warren G (1995) Reassembly of Golgi stacks from mitotic Golgi fragments in a cell-free system. J Cell Biol 129:605–618

23. Misteli T, Warren G (1994) COP-coated vesicles are involved in the mitotic fragmentation of Golgi stacks in a cell-free system. J Cell Biol 125:269–282

24. Tang D, Xiang Y, Wang Y (2010) Reconstitution of the cell cycle-regulated Golgi disassembly and reassembly in a cell-free system. Nat Protoc 5:758–772

25. Tang D, Mar K, Warren G, Wang Y (2008) Molecular mechanism of mitotic Golgi

disassembly and reassembly revealed by a defined reconstitution assay. J Biol Chem 283:6085–6094

26. Munro S (2011) The golgin coiled-coil proteins of the Golgi apparatus. Cold Spring Harb Perspect Biol 3:1–14

27. Feinstein TN, Linstedt AD (2008) GRASP55 regulates Golgi ribbon formation. Mol Biol Cell 19:2696–2707

28. Puthenveedu MA, Bachert C, Puri S, Lanni F, Linstedt AD (2006) GM130 and GRASP65-dependent lateral cisternal fusion allows uniform Golgi-enzyme distribution. Nat Cell Biol 8:238–248

29. Thyberg J, Moskalewski S (1999) Role of microtubules in the organization of the Golgi complex. Exp Cell Res 246:263–279

30. Rios RM, Bornens M (2003) The Golgi apparatus at the cell centre. Curr Opin Cell Biol 15:60–66

31. Marra P, Salvatore L, Mironov A Jr, Di Campli A, Di Tullio G, Trucco A, Beznoussenko G, Mironov A, De Matteis MA (2007) The biogenesis of the Golgi ribbon: the roles of membrane input from the ER and of GM130. Mol Biol Cell 18:1595–1608

32. Jarvela T, Linstedt AD (2014) Isoform-specific tethering links the Golgi ribbon to maintain compartmentalization. Mol Biol Cell 25:133–144

33. Yin DX, Schimke RT (1995) BCL-2 expression delays drug-induced apoptosis but does not increase clonogenic survival after drug treatment in HeLa cells. Cancer Res 55:4922–4928

Chapter 15

The Role of Lysophospholipid Acyltransferases in the Golgi Complex

John A. Schmidt

Abstract

Determining the abundance of phospholipids and neutral lipids in cellular membranes is paramount to understanding their biological functions. Many lipid-modifying enzymes have yet to be characterized due to limitations in substrate–product measurements and purification of membrane-bound enzymes. The method described here uses radiolabeled phospholipid substrates and cell-purified organelles to quantify phospholipid metabolism using thin-layer chromatography. This assay has the benefits of being specific and adaptable for numerous applications and systems.

Key words Phosphatidic acid, Lysophosphatidic acid, LPAT, LPAAT, Acyltransferase, Golgi membranes

1 Introduction

Organelle phospholipid species have roles in cellular signaling, protein docking, raft formation, and morphology [1–4]. New understanding of the physiological roles of phospholipids in organelle membranes requires experimental methods to measure specific phospholipid metabolism in an environment with vast phospholipid diversity. However, elucidating the roles of phospholipids in cellular functions has been limited by our abilities to manipulate and quantify them. This chapter describes methods that were developed to measure two phospholipids that are important for the structure and function of the Golgi complex—lysophosphatidic acid (LPA) and phosphatidic acid (PA). Both LPA and PA have been shown to be important for Golgi membrane trafficking by altering membrane shape and transport vessel biogenesis [5, 6]. Moreover, PA is a centrally important phospholipid because it is a metabolic precursor for all membrane phospholipids and many neutral lipids including diacylglycerol [2, 7].

PA and LPA levels are in part controlled by enzymes that reacylate LPA to PA—lysophosphatidic acid acyltransferase (LPAAT) (Fig. 1a). Indeed, one such enzyme, human LPAAT3 (AGPAT3)

William J. Brown (ed.), *The Golgi Complex: Methods and Protocols*, Methods in Molecular Biology, vol. 1496, DOI 10.1007/978-1-4939-6463-5_15, © Springer Science+Business Media New York 2016

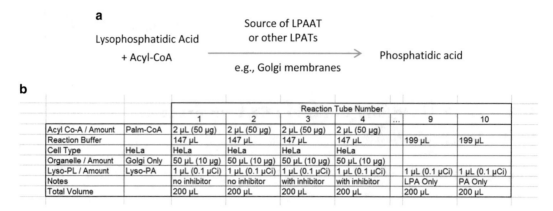

a

Lysophosphatidic Acid

+ Acyl-CoA

Source of LPAAT
or other LPATs

e.g., Golgi membranes

→ Phosphatidic acid

b

		Reaction Tube Number						
		1	2	3	4	...	9	10
Acyl Co-A / Amount	Palm-CoA	2 µL (50 µg)	2 µL (50 µg)	2 µL (50 µg)	2 µL (50 µg)			
Reaction Buffer		147 µL	147 µL	147 µL	147 µL		199 µL	199 µL
Cell Type	HeLa	HeLa	HeLa	HeLa	HeLa			
Organelle / Amount	Golgi Only	50 µL (10 µg)	50 µL (10 µg)	50 µL (10 µg)	50 µL (10 µg)			
Lyso-PL / Amount	Lyso-PA	1 µL (0.1 µCi)	1 µL (0.1 µCi)	1 µL (0.1 µCi)	1 µL (0.1 µCi)		1 µL (0.1 µCi)	1 µL (0.1 µCi)
Notes		no inhibitor	no inhibitor	with inhibitor	with inhibitor		LPA Only	PA Only
Total Volume		200 µL	200 µL	200 µL	200 µL		200 µL	200 µL

Fig. 1 (**a**) LPAAT reaction. (**b**) Sample table for setting up a series of LPAAT reactions

has been shown to regulate Golgi membrane tubule and COPI vesicle formation [1, 6]. In vitro LPAAT assays can be used to measure changes in LPA and PA in organelle membranes using radiolabeled substrates separated by thin-layer chromatography (TLC) [5, 6]. This assay is highly sensitive and specific, but also flexible, and can be modified for numerous other applications.

TLC has been a proven method for the separation of phospholipids for decades [8, 9]. Recent advances in lipidomics use liquid chromatography and mass spectrometry (LC-MS) to measure molecular lipid species [10, 11]. While these methods are useful for obtaining global information about phospholipid levels, there are many drawbacks. MS can have a large range of ionization efficiency that is complicated by the thousands of headgroups and acyl chain combinations that make up phospholipids [12]. Furthermore, such approaches are not useful for kinetic analyses or to test numerous samples. Additionally, phospholipid metabolic pathways are complex and interrelated, so a TLC-based analysis is often required to validate enzyme activity observed in lipidomic studies [7].

Measuring the enzymatic activity of transmembrane proteins that modify phospholipids is particularly challenging because of the difficulty of delivering substrates, quantifying products, and purifying the enzyme. The human genome encodes several integral membrane LPAAT enzymes, in addition to numerous other LPATs [13, 14]. This LPAAT assay allows for the manipulation of protein expression and sequences using molecular biology in a eukaryotic cellular environment and for the sensitive quantification of phospholipid products, which could be applied for the study of any LPAT enzyme.

2 Materials

Many of these reagents contain organic solvents or strong acids and should be made, and stored, in a well-ventilated fume hood with protective goggles, nitrile gloves, protective coat, and/or face shield. Due to evaporation rates, organic solvents should be made fresh for each use. Measuring and transfer of organic solvents and acids should be done with chloroform-rinsed glass syringes or graduated cylinders. Radioactive substrates should be handled with care per your institution's recommendations for use.

1. Chromatography solvent: chloroform–methanol–acetic acid–water (60:50:1:4 v/v ratio). It is extremely important to use chromatography grade reagents for well-defined and consistent separations. Store at room temperature in a fume hood.

2. Reaction buffer: 10 mM Tris base pH 7.4, 150 mM NaCl, 1 mM EDTA. Dissolve 1.21 g of Tris base, 8.77 g NaCl, and 0.292 g EDTA (ethylenediaminetetraacetic acid) in 900 mL distilled water. Use 1 M HCl to bring the pH to 7.4 and bring final volume to 1 L. Store at 4 °C.

3. Stop buffer: chloroform–methanol–water (316:632:53 v/v ratio) with chromatography grade reagents. Store at room temperature in a fume hood.

4. Spotting buffer: chloroform–methanol (1:1 ratio).

5. Copper (II) sulfate stain: 3% $CuSO_4$ (w/v) and 8% H_3PO_4 (v/v). Dissolve 6 g of $CuSO_4$ in 184 mL of distilled water and slowly add 16 mL of H_3PO_4 phosphoric acid (HPLC grade 85–90%).

6. Lysophospholipids and phospholipids.

7. Acyl-CoA fatty acids: palmitoyl Coenzyme A (ammonium salt) or 16:0 CoA.

8. Tritium radiolabeled lysophospholipids: lysophosphatidic acid, 1-Oleoyl-[Oleoyl-9,10-^{3}H]-, 0.1 mCi/mL from PerkinElmer (Waltham, MA).

9. Thin-layer chromatography (TLC) plates: glass backing with silica gel, polymer or organic binder, and no fluorescent indicator 10 cm^2.

10. Drying oven set to 110–120 °C.

11. Rectangular thin-layer chromatography developing tank with lid.

12. Argon gas tank with regulator and adjustable valve.

13. Corning® Pyrex® Screw Cap Culture Tubes with PTFE Lined Phenolic Caps.

14. PTFE tape.

15. Bath sonicator.

16. Water bath at 37 °C with test tube rack.

17. Vortex mixer.

18. Clinical centrifuge.

19. Imager Typhoon FLA 7000 (GE Healthcare Lifesciences) or similar.

20. Storage Phosphor screen and cassette (GE Healthcare Lifesciences).

21. Reagent sprayer.

22. Capillary tubes.

23. Graduated glass syringes with 2 μL–1000 μL range.

24. Golgi membranes isolated from mammalian cells grown in culture as described [15] or from dissected liver [16] (*see* **Note 1**).

3 Methods

3.1 Preparation

1. Rinse thin-layer chromatography developing tank with chromatography solvent and add solvent to the tank at an approximate depth of 0.5 cm. Place the lid on the tank, with a weight if needed, and allow the vapors in the tank to equilibrate overnight at room temperature in ventilated safety cabinet.

2. Prepare TLC plates by washing them in chromatography solvent in a glass dish. Place the washed TLC plates in a drying oven at 110 °C until ready to use.

3. Dissolve acyl-CoA fatty acids and nonradioactive LPA in chloroform–methanol 1:1 to a concentration of 25 mg/mL. Place stock solution in glass vials with PTFE caps and gently replace the atmospheric gas in the vial with argon gas (or nitrogen gas) to minimize oxidation (*see* **Note 2**). Vial caps should be firmly attached and secured with Teflon tape or Parafilm. Stock solutions may be stored for up to 1 year at –20 °C (*see* **Note 3**).

3.2 LPAAT Reaction

1. Begin by planning your reactions and determining the amount of each reagent needed (Fig. 1b). Allow lipids and buffer to warm to room temperature. For each LPAAT reaction, add 1 μL of radiolabeled lysophosphatidic acid, 1-Oleoyl-[Oleoyl-9,10-^{3}H]- or 0.1 μCi to a glass vial (*see* **Note 4**). Dry the solvent under a gentle stream of argon gas—a glassy-white film may be seen on the bottom of the tube.

2. For each reaction, add 25 μL of assay buffer to the dried lipids. Cap the vial and dissolve the lipids in solution with a few 3–5 s. pulses on a vortex mixer. The mixture is usually slightly cloudy.

3. Place the vial in a bath sonicator with sufficient water to submerge the lipid mixture. Do not allow the glass vial to contact the metal basin of the sonicator. Turn on the sonicator with the immersed reaction vial for 7 min. Sonication should be done in the dark. Following sonication the mixture should appear clear and produce small bubbles if flicked with your fingers. Store the radioactive lipid mixture in the dark until the reactions are set up.

4. To plan your reactions consider the origin of the Golgi membranes being tested. If the Golgi were isolated from an animal, record the species, age, gender, and genetic background. If the Golgi were isolated from cultured cells, the cell line, the cell origin, and transfected constructs (genes/shRNA/siRNA) should be considered. Thaw enough Golgi membrane for the assay on ice. Each reaction should be done in duplicate with 10 μg Golgi protein per reaction.

5. Prepare each reaction in a labeled glass vial. Add 10 μg of isolated Golgi membranes and enough reaction buffer for a total volume of 173 μL. Transfer Golgi membranes with a pipet tip that has a larger hole to prevent rupturing the membranes. Add 2 μL of palmitoyl-CoA (*see* **Note 5**). Lastly, add 25 μL of the prepared radiolabeled LPA that has been stored in the dark.

6. Cap the reaction tubes, wrap the caps with Parafilm, and gently mix by swirling the tubes. Place the reaction tubes in a 37 °C water bath for the duration of the reaction, or 1 h (*see* **Note 6**).

3.3 Extraction

1. Stop the reaction by adding 1.0 mL of stop buffer for a final v/v ratio of chloroform–methanol–water 1:2:0.8 (*see* **Note 7**).

2. Replace the cap and vortex extensively while gripping the center of the tube to minimize splashing on the tube walls. Allow the reactions to rest at room temperature for 5 min.

3. Vortex the reactions again, then add 300 μL of chloroform and 300 μL of water to each reaction tube. Vortex again and allow the tubes to rest at room temperature for 5 min.

4. Centrifuge the reaction tubes at low speed (about $1600 \times g$ in a clinical centrifuge) for 5 min with rubber inserts to prevent glass tubes from breaking.

5. The reaction will have separated into two phases—a lower hydrophobic phase and an upper hydrophilic phase. With a syringe, carefully remove the lower phase, which should be 600 μL in volume, and place it in a clean glass reaction tube.

6. Once the lower phase from each reaction has been transferred, "spike" each mixture with 10 μg of nonradioactive LPA and 10 μg of PA. These additions will help identify the radiolabeled reaction products and substrates.

7. Evaporate each of the organic phases by gently placing them under a slow stream of argon gas. Be cautious not to splash the products on the sides of the tubes (*see* **Note 8**). When completely dry, a white film can be seen at the rounded bottom of the tube.

3.4 Separation

1. Replace the solvent in the thin-layer chromatography developing tank with fresh solution so that the solvent is about 0.5–1.0 cm from the bottom of the tank (*see* **Notes 9** and **10**). Replace the lid of the tank.

2. Remove a silica thin-layer chromatography plate from the drying oven and use a pencil to draw a horizontal line 2 cm from the bottom edge. Be gentle with the pencil or the silica will chip off the plate. From left to right, place a short vertical pencil line every 1 cm to the edge. Each pencil cross is a guide for where to spot the phospholipid mixture. Label each cross with a number just below the pencil mark intersections.

3. To spot the lipids onto the plate, dissolve the dried lipids in 25 µL of chloroform/methanol 1:1. Spot the mixture onto the plate where the pencil crosshairs can be seen. Use a capillary or syringe to slowly add the mixture so the circumference of the wet mark on the plate does not exceed 3–5 mm. When one droplet dries, more of the lipids can be added to the spot until the entire 25 µL has been spotted and dried. Repeat this for each reaction.

4. It is recommended that a lane for nonradioactive PA only and LPA only be included as controls.

5. When all of the reactions have been spotted and the plate is completely dry, gently place the plate in the developing tank such that the plate is vertical and the bottom of the plate forms an angle towards the front and the top of the plate leans away. Most developing tanks have a ridge on the bottom of the tank to support the plate.

6. The solvent will push the samples up the plate and separate them. When the solvent front reaches the top of the plate, the separation is finished and remove the plate from the tank. This usually takes about 15–20 min for a 10 cm plate.

7. Allow the TLC plate to completely dry of all solvents at room temperature.

3.5 Detection

1. Prepare a phosphor screen by removing any residual background signal. Exposure to bright light for 10 min is sufficient.

2. Wrap the dry TLC plate in plastic wrap to protect the screen. Put the plate in a storage cassette and place the phosphor screen on top. Keep the cassette and screen in the dark until they are ready to scan. Exposure time can vary depending on how robust the reactions were. Try a short exposure time of 1 h followed by longer times up to 24 h if needed.

3. To scan the phosphor screen, quickly remove the screen and place it on the scanner to minimize bleaching from room lights. Follow the manufacturer's instructions to detect tritium radioactivity (Fig. 2).

4. When a satisfactory image is obtained, the plate can then be stained for total lipids. Either purchase or construct a cardboard box that can hold the TLC plate in a vertical position,

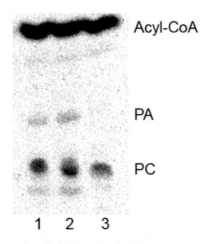

Fig. 2 Production of radiolabeled phosphatidic acid and phosphatidylcholine. An LPAAT Assay results using Golgi and microsomes from HeLa cells overexpressing LPAATγ (AGPAT3) in *Lane 1*, LPAATζ (AGPAT6) in *Lane 2*, or empty vector in *Lane 3*

but has one side open. Add the copper sulfate stain to the sprayer and coat the plate with a thin layer of the solution (*see* **Notes 10** and **11**).

5. Allow the plate to dry in the drying oven. This will char the plate and turn spots of lipids light brown to dark brown. A higher temperature can be used to visualize more lipids. Oven temperatures can vary from 110 to 180 °C. Hotter temperatures will be quicker. If the plates are left in the oven too long, they will become too dark. Typically, 20 min is sufficient to visualize the most important phospholipids in this assay (*see* Fig. 3).

6. The stained plate can be cooled and digitally scanned or measured to correlate stained sports with radioactive signals to differentiate between radiolabeled substrates and products. Including spots on the TLC plate that are only LPL and only palmitoyl-CoA may be helpful. Digital densitometry measurements to quantify products is recommended.

7. Compare the generation of radiolabeled PA and the loss of radiolabeled LPA to controls.

8. Radioactive waste should be disposed of based on your institution's guidelines. All equipment should be monitored for radioactivity contamination and cleaned as needed.

4 Notes

1. We have used isolated Golgi membranes and endoplasmic reticulum microsomes with this assay, but other membrane bound organelles may also substituted. Additionally, a post-nuclear lysate may also be used, but amount needed for robust results may need to be tested.

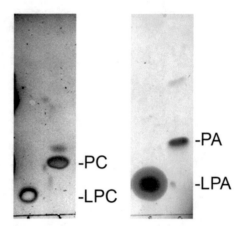

Fig. 3 Copper sulfate stained TLC plates. Mobility of phosphatidic acid (PA), lyso-phosphatidic acid (LPA), phosphatidylcholine (PC), and lyso-phosphatidylcholine (LPC) stained with copper sulfate and oven-charred for 20 min

2. Nitrogen gas or other nonreactive gases may be substituted.

3. If mobility of phospholipids changes or multiple spots are seen when only one is expected, reagents may have started to degrade or oxidize and should be replaced.

4. Different lysophospholipid substrates may be used instead including lyso-phosphatidylcholine, lyso-phosphatidylethanolamine, lyso-phosphatidylglycerol, and lyso-phosphatidylserine.

5. While palmitoyl-CoA is a physiologically relevant fatty acid for many organelle membranes, the procedure can be adjusted to test different fatty acids of different lengths or saturation. For this alternative, the radiolabel should be placed on the acyl-CoA instead of the lysophospholipid to determine which acyl-CoA is preferred.

6. The time of the reaction can be varied if the reaction products are robust. A 1 h reaction is appropriate for LPAAT assays. Shorter reactions may be needed for more abundant phospholipids.

7. Extraction methods are based on Bligh & Dyer method [17].

8. For experiments with many samples, a commercial multi-tube evaporator may be useful (e.g., N-Evap 111, Organomation Associates). We recommend this to save time on large experiments.

9. If detection of other phospholipids is required, the solvent can be changed to adjust the mobility in the silica plate. Specifically, separation of phosphoinositides and phosphatidylethanolamine may need a different solvent [18].

10. Acetic acid is important for the proper separation of PA and other charged phospholipids.

11. Alternate stains are available for specific visualization, we chose this solution for its ease and cost. Copper sulfate stain can also be reused.

12. If a sprayer is not available, quickly dipping the TLC plate in the copper sulfate solution can also work (personal communication E. Cluett).

Acknowledgements

Thank you William J. Brown for critical reading of the manuscript and Edward Cluett, Dan Drecktrah, and Kimberly Chambers for their intellectual contributions.

References

1. Yang JS, Valente C, Polishchuk RS, Turacchio G, Layre E, Moody DB, Leslie CC, Gelb MH, Brown WJ, Corda D, Luini A, Hsu VW (2011) COPI acts in both vesicular and tubular transport. Nat Cell Biol 13:996–1003

2. Ha KD, Clarke BA, Brown WJ (2012) Regulation of the Golgi complex by phospholipid remodeling enzymes. Biochim Biophys Acta 1821:1078–1088

3. Bankaitis VA, Garcia-Mata R, Mousley CJ (2012) Golgi membrane dynamics and lipid metabolism. Curr Biol 22:R414–R424

4. Graham TR, Kozlov MM (2010) Interplay of proteins and lipids in generating membrane curvature. Curr Opin Cell Biol 22:430–436

5. Drecktrah D, Chambers K, Racoosin EL, Cluett EB, Gucwa A, Jackson B, Brown WJ (2003) Inhibition of a Golgi complex lysophospholipid acyltransferase induces membrane tubule formation and retrograde trafficking. Mol Biol Cell 14:3459–3469

6. Schmidt JA, Brown WJ (2009) Lysophosphatidic acid acyltransferase 3 regulates Golgi complex structure and function. J Cell Biol 186:211–218

7. Shui G, Guan XL, Gopalakrishnan P, Xue Y, Goh JS, Yang H, Wenk MR (2010) Characterization of substrate preference for Slc1p and Cst26p in Saccharomyces cerevisiae using lipidomic approaches and an LPAAT activity assay. PLoS One 5:e11956

8. Skipski VP, Peterson RF, Barclay M (1964) Quantitative analysis of phospholipids by thin-layer chromatography. Biochem J 90:374–378

9. Seminariodebohner L, Soto EF, Decohan T (1965) Quantitative analysis of phospholipids by thin-layer chromatography. J Chromatogr 17:513–519

10. Duran JM, Campelo F, van Galen J, Sachsenheimer T, Sot J, Egorov MV, Rentero C, Enrich C, Polishchuk RS, Goni FM, Brugger B, Wieland F, Malhotra V (2012) Sphingomyelin organization is required for vesicle biogenesis at the Golgi complex. EMBO J 31:4535–4546

11. Dennis EA (2009) Lipidomics joins the omics evolution. Proc Natl Acad Sci U S A 106:2089–2090

12. Myers DS, Ivanova PT, Milne SB, Brown HA (2011) Quantitative analysis of glycerophospholipids by LC-MS: acquisition, data handling, and interpretation. Biochim Biophys Acta 1811:748–757

13. Leung DW (2001) The structure and functions of human lysophosphatidic acid acyltransferases. Front Biosci 6:D944–D953

14. Yamashita A, Hayashi Y, Matsumoto N, Nemoto-Sasaki Y, Oka S, Tanikawa T, Sugiura T (2014) Glycerophosphate/acylglycerophosphate acyltransferases. Biology (Basel) 3:801–830

15. Yang JS, Gad H, Lee SY, Mironov A, Zhang L, Beznoussenko GV, Valente C, Turacchio G, Bonsra AN, Du G, Baldanzi G, Graziani A, Bourgoin S, Frohman MA, Luini A, Hsu VW (2008) A role for phosphatidic acid in COPI vesicle fission yields insights into Golgi maintenance. Nat Cell Biol 10:1146–1153

16. Cluett EB, Brown WJ (1992) Adhesion of Golgi cisternae by proteinaceous interactions: intercisternal bridges as putative adhesive structures. J Cell Sci 103:773–784

17. Bligh EG, Dyer WJ (1959) A rapid method of total lipid extraction and purification. Can J Biochem Physiol 37:911–917

18. Chambers K, Brown WJ (2004) Characterization of a novel CI-976-sensitive lysophospholipid acyltransferase that is associated with the Golgi complex. Biochem Biophys Res Commun 313:681–686

Chapter 16

Methods to Purify and Assay Secretory Pathway Kinases

Vincent S. Tagliabracci, Jianzhong Wen, and Junyu Xiao

Abstract

Members of the four-jointed and VLK families of secretory pathway kinases appear to be responsible for the phosphorylation of secreted proteins and proteoglycans. These enzymes have been implicated in many biological processes and mutations in several of these kinases cause human diseases. Here, we describe methods to purify and assay two members of the four-jointed family of secretory kinases: the Fam20C protein kinase and the Fam20B proteoglycan kinase.

Key words Raine syndrome, Hypophosphatemia, Golgi kinase, Secreted kinase, Amelogenesis imperfecta

1 Introduction

Protein kinases are evolutionarily conserved enzymes that covalently modify substrates with a molecule of phosphate, in most instances using ATP as the phosphate donor [1]. The phosphoacceptor substrate is usually specific for a given protein kinase and includes, but is not limited to, Ser, Thr and Tyr residues on proteins, inositol head groups on lipids, and hydroxyl groups on sugars. Since the discovery of the first protein kinase [2], most kinase research has focused on the phosphorylation of intracellular substrates because the "lion's share" of protein kinases localize within the boundaries of the plasma membrane [3]. Nevertheless, proteins and proteoglycans that are secreted can be extensively modified by phosphorylation as well [4–6]. For example, the secreted milk proteins, casein and osteopontin, are phosphorylated on multiple Ser residues [7, 8]. Furthermore, proteoglycans, such as decorin, contain phosphate within the tetrasaccharide linkage region and this phosphate is covalently linked to the 2′ OH on a xylose residue [9, 10]. However, efforts to determine the functional consequences of secreted protein and proteoglycan phosphorylation have been hampered because the kinases responsible for these phosphorylation events have only recently been identified

William J. Brown (ed.), *The Golgi Complex: Methods and Protocols*, Methods in Molecular Biology, vol. 1496,
DOI 10.1007/978-1-4939-6463-5_16, © Springer Science+Business Media New York 2016

[5, 11]. In 2008, Irvine and colleagues discovered that the *Drosophila* protein four-jointed (fj), is a Golgi kinase that phosphorylates atypical cadherins [12]. This study revealed for the first time the molecular identity of a kinase in the secretory pathway. Subsequent studies identified Fam20B as a xylose kinase that plays a role in regulating proteoglycan biosynthesis [13]. Fam20B dependent phosphorylation of xylose markedly stimulates the activity of galactosyltransferase II (GalT-II) to enhance proteoglycan biosynthesis [14]. In 2012, we discovered that Fam20C is the bona fide "Golgi casein kinase," an enzyme that escaped identification for many years [15]. Fam20C phosphorylates hundreds of secreted proteins on Ser-x-Glu/pSer motifs and appears to be responsible for generating the vast majority of the extracellular phosphoproteome [5, 15–18].

Although human mutations in Fam20B and Four-jointed have yet to be identified, Fam20C mutations cause Raine syndrome, an often-fatal osteosclerotic bone dysplasia [19, 20]. Most Raine patients die within the first few weeks of life; however, more recent reports have identified non-lethal cases of the disease in patients harboring hypomorphic *FAM20C* mutations [17, 21, 22]. These individuals, as well as Fam20C KO mice, have dental abnormalities and hypophosphatemia, the later of which is caused by elevated levels of the phosphate-regulating hormone FGF23 [21, 23, 24]. Fam20A, a close paralogue of Fam20C, is a pseudokinase that forms a functional complex with Fam20C to enhance the phosphorylation of secreted proteins that are found in the enamel matrix of the tooth [25]. Human mutations in Fam20A cause amelogenesis imperfect and enamel–renal syndrome, which results in severe enamel defects in patients [26, 27].

Recent work from the Whitman laboratory has identified the vertebrate lonesome kinase (Vlk) as a secreted kinase that phosphorylates tyrosine residues [28]. VLK belongs to the PKDCC family of kinases that are distantly related to a protein kinase family consisting of Fam69A/B/C and Deleted in Autism-1 (DIA1) and DIA1-Related [29]. These proteins localize in the secretory pathway and some of them have been genetically linked to neurological disorders [16, 30]. VLK phosphorylates a wide range of extracellular proteins, including matrix metalloproteinase 1 and ERp29 with no recognizable consensus sequence surrounding the phosphoacceptor Tyr [28].

Moreover, Sgk196, which was originally annotated as pseudokinase, is a sugar kinase that phosphorylates mannose residues on α-dystroglycan [31]. SGK196 was discovered in a haploid screen as a gene involved in the post-translational modification of α-dystroglycan, and human mutations in *SGK196* lead to α-dystroglycan-related diseases such as congenital muscular dystrophy [32, 33].

Here, we describe methods for the purification of milligram quantities of Fam20C and Fam20B from insect cell conditioned medium using recombinant baculovirus. The proteins can be used for many applications including X-ray crystallography and enzyme assays. We also describe methods to assay Fam20C and Fam20B activities in vitro and in cells. We hope that the methods described herein will encourage further research on secreted protein and proteoglycan phosphorylation by kinases in the secretory pathway and extracellular space.

2   Materials

Prepare all solutions using ultrapure water (Millipore) and analytical grade reagents. Prepare and store all reagents at room temperature (unless indicated otherwise). Diligently follow all waste disposal regulations when disposing waste materials. The $[\gamma\text{-}^{32}P]$ATP and $\gamma\text{-}^{32}P$-orthophosphate must be handled in accordance with regulations for the use of radioactivity. Experiments should be performed behind a plexiglass shield to minimize exposure to radiation. A lab coat, gloves, and safety glasses should be worn at all times during the procedure. The radioactive samples and the source of radioactivity must be shielded at all times. The stock vials should be stored in a lead case or a plexiglass box. The time spent handling radioactive samples should be minimized to avoid prolonged exposure.

2.1   Generation of Recombinant Fam20B and Fam20C Using Insect Cells

1. Baculovirus transfer vector pI-sec-MBP-tev2 (pSMBP2 as an abbreviation, Fig. 1). The pSMBP2 vector is constructed on the basis of pI-secSUMOstar (LifeSensors) and pFastBac™HT-B (Invitrogen). It contains a signal peptide (SP) sequence from the baculovirus glycoprotein gp67, followed by regions encoding a 6X His tag, a maltose-binding protein (MBP), a 7-residue spacer, and a Tobacco Etch Virus (TEV) protease cleavage site (Fig. 1).

2. MAX Efficiency DH10Bac *E. coli* competent cells (10361-012, Invitrogen). Growth plate: LB agar plate containing 50 μg/mL kanamycin, 10 μg/mL tetracycline, 7 μg/mL gentamicin, 40 μg/mL IPTG, and 100 μg/mL Bluo-gal. Growth medium: LB medium supplemented with kanamycin, tetracycline, and gentamicin at above concentrations.

3. PureLink HiPure Plasmid DNA miniprep Kit.

4. Sf9 and Sf21 insect cells (Invitrogen). Growth medium: Sf-900 II SFM (Invitrogen).

5. Transfection reagent: Cellfectin or X-tremeGENE 9.

6. High Five (Tn5) insect cell. Growth medium: ESF 921 (Expression Systems).

7. TBS buffer: 25 mM Tris pH 7.5, 150 mM NaCl.

8. Ni-NTA (nickel nitrilotriacetic acid) agarose. Wash buffer: 25 mM Tris pH 8.0, 150 mM NaCl, 25 mM imidazole. Elution buffer: 25 mM Tris pH 8.0, 150 mM NaCl, 250 mM imidazole.

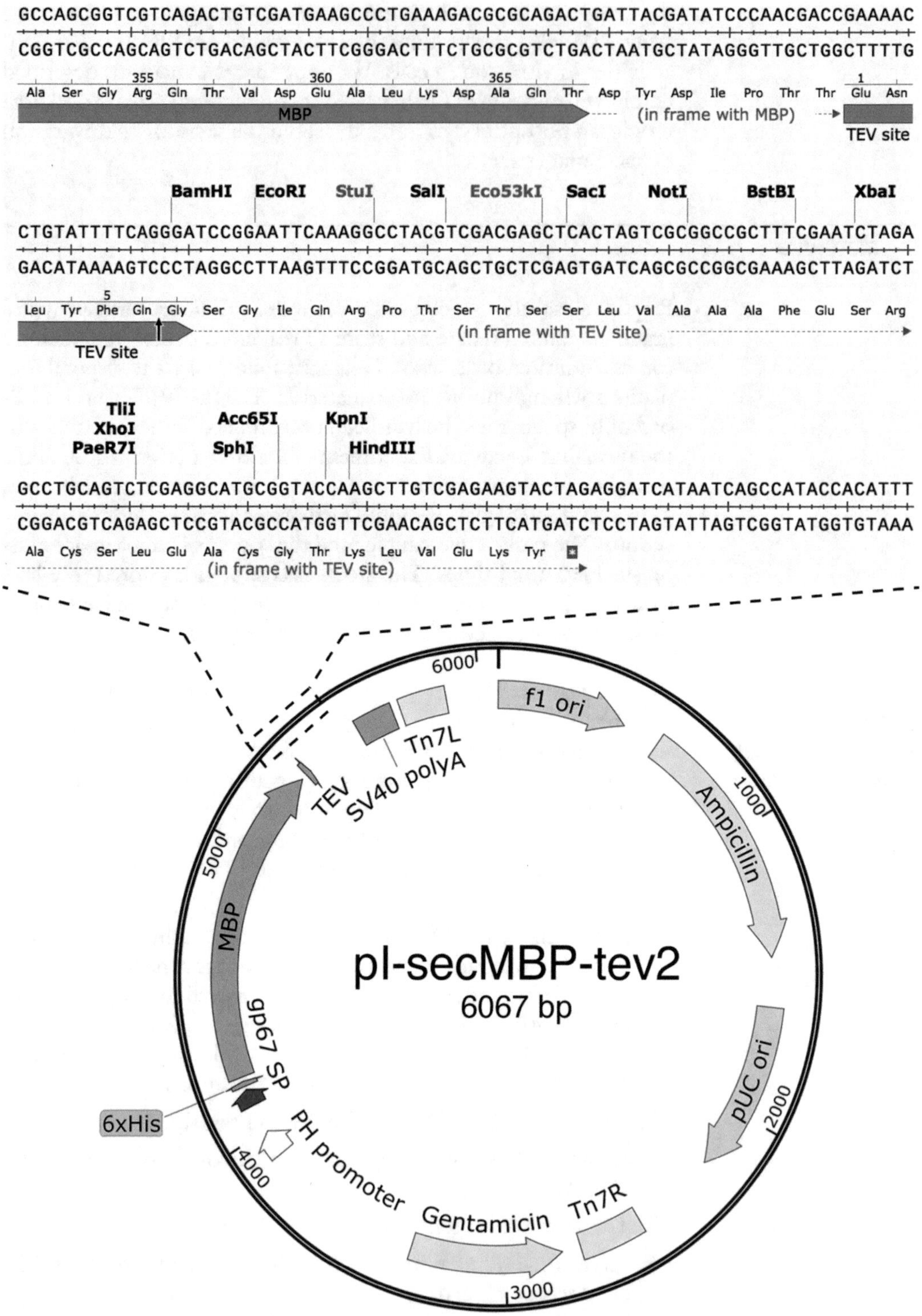

Fig. 1 pSMBP2 Vector Map. Schematic of the pSMBP2 Baculovirus transfer vector depicting the signal peptide (SP) sequence from the baculovirus glycoprotein gp67, the 6X His tag, the maltose-binding protein (MBP), a 7-residue spacer, and a Tobacco Etch Virus (TEV) protease cleavage site. The multiple cloning site is also shown

9. TEV protease.

10. Resource Q column (GE Life Sciences). Buffer A: 25 mM Tris pH 8.0, 50 mM NaCl; Buffer B: 25 mM Tris pH 8.0, 1 M NaCl.

11. Hiload Superdex 200 column (GE Life Sciences). Running buffer: 10 mM HEPES pH 7.5, 100 mM NaCl.

12. Equipment: refrigerated incubator and shaker, hemocytometer, light microscope, cross-flow filtration system (Vivaflow 200, Sartorius), fast protein liquid chromatography system.

2.2 Fam20B Sugar Kinase Assay Using a Model Substrate

1. Model sugar substrate: Glucuronic acid-β1-3-Galactose-β1-3-Galactose-β1-4-Xylose-β1-Benzyl (Tetra-Ben) (customized from TCI America, Portland, OR, USA). Dissolve 1 mg Tetra-Ben in 135 μL water to make a 10 mM stock solution (Tetra-Ben molecular weight: 740 g/mol). Store aliquots at –20 °C.

2. Purified recombinant Fam20B wild type and D309A inactive mutant protein: 1 mg/mL in TBS buffer and 20% glycerol, aliquot and store at –80 °C.

3. 5× Fam20B kinase buffer: 250 mM HEPES, pH 7.4, 50 mM MgCl$_2$, 500 mM NaCl, 1% Triton X-100.

4. MnCl$_2$.

5. Adenosine 5′-triphosphate disodium salt hydrate (ATP).

6. [γ-^{32}P]ATP.

7. Kinase reaction quenching buffer: 0.1 M EDTA, 2 mM ATP.

8. Sep-Pak C18 Vac RC Cartridge, 100 mg (Waters, Milford, MA, USA).

9. Cartridge activation/elution solvent: methanol.

10. Cartridge equilibration and wash buffer: 0.2 M (NH$_4$)$_2$SO$_4$.

11. Equipment: Liquid scintillation spectrometry (scintillation counter); liquid scintillation counting cocktails; liquid scintillation counting vials.

2.3 Fam20B Proteoglycan Kinase Assay Using Decorin (DCN) as a Substrate

1. Plasmid: DCN (1-359) with C-terminal FLAG tag in pCCF vector (pcDNA backbone containing a C-terminal Flag tag).

2. HEK293T cell line.

3. Tissue culture components: Nunc TripleFlasks (Sigma), Gibco® DMEM cell growth medium, Fetal Bovine Serum (FBS), DPBS, Pen/Strep.

4. Transfection reagent: linear polyethylenimine (PEI, molecular weight: 25 kDa). Dissolve PEI at 1 mg/mL in H$_2$O preheated to 80 °C. Cool solution to RT and adjust pH to 7.2. Filter-sterilize, aliquot and store at –20 °C.

5. Anti-FLAG M2 affinity gel; 3X-FLAG peptide.

6. Empty PD-10 column (GE Healthcare Life Sciences, Pittsburg, PA, USA).

7. Amicon Ultra-4 centrifugal filter units (EMD Millipore, Billerica, MA, USA).

8. Recombinant DCN from mammalian cells: 1 mg/mL. Store at −20 °C.

9. Glycosaminoglycan (GAG) treatment enzymes: Heparinase I, II, and III (Flavobacterium heparinum, Seikagaku, Japan), chondroitinase ABC (Chon-ABC). Store at −20 °C.

10. 10× GAG treatment buffer: 500 mM sodium acetate, 5 mM calcium acetate, pH 7.0. Store at −20 °C.

11. 5× Fam20B kinase buffer: 250 mM HEPES, pH 7.4, 50 mM $MgCl_2$, 500 mM NaCl, 1 % Triton X-100.

12. $MnCl_2$.

13. Adenosine 5′-triphosphate disodium salt hydrate (ATP).

14. $[\gamma\text{-}^{32}P]$ATP.

15. Kinase reaction quenching buffer: 0.1 M EDTA and 2 mM ATP.

16. Gel electrophoresis systems.

17. 5× SDS-PAGE loading buffer: 62.5 mM $Tris\text{-}PO_4$ pH = 6.8, 50 % (w/v) glycerol, 6.25 % SDS, 0.1 % bromophenol blue, 5 % β-mercaptoethanol.

2.4 Fam20B Activity in Cells

1. Tissue culture components: MRC-5 lung fibroblast control and stable knockdown cell lines, six-well cell culture plates, Dialyzed FBS, Phosphate free DMEM.

2. $[^{32}P]$orthophosphate, 10 mCi/mL.

3. Protein G Agarose.

4. Complete protease inhibitor (PIC).

5. Phosphatase inhibitor cocktail.

6. Gel electrophoresis and transfer system.

7. Anti-DCN antibody (MAB143, R&D systems, Minneapolis, MN, USA).

2.5 Fam20C Activity Using β(28-40) as the Substrate

1. Insect cell purified Fam20C (63-C).

2. β(28-40) peptide KKIEKFQSEEQQQ (commercially available from many sources).

3. 10× Fam20C kinase Buffer: 500 mM HEPES–KOH pH 7.0, 5 mg/mL bovine serum albumin (BSA).

4. $MnCl_2$.

5. Adenosine 5′-triphosphate disodium salt hydrate.

6. $[\gamma\text{-}^{32}P]$ATP.

7. P81 phosphocellulose filters (GE Healthcare). This product has been discontinued. A suitable replacement is P81 phosphocellulose filters (20-134, Millipore).

8. 85 % phosphoric acid.

9. Acetone.

10. Equipment: water bath, wire mesh basket inside of a beaker with a stir bar, stir plate, liquid scintillation spectrometry (scintillation counter); liquid scintillation counting cocktails; liquid scintillation counting vials.

2.6 Fam20C Activity Using Recombinant OPN as the Substrate

1. Insect cell purified Fam20C (63-C).

2. Recombinant 6X-His-tagged human osteopontin (residues 17-314) expressed and purified from *E. coli* using standard procedures.

3. HEPES.

4. $MnCl_2$.

5. Gel electrophoresis systems.

6. EDTA.

7. 5× SDS-PAGE loading buffer: 62.5 mM Tris-PO_4 pH 6.8, 50% (w/v) glycerol, 6.25% SDS, 0.1% bromophenol blue, β-mercaptoethanol.

8. Adenosine 5′-triphosphate disodium salt hydrate (ATP).

9. [γ-^{32}P]ATP.

10. Equipment: water bath, liquid scintillation spectrometry (scintillation counter); liquid scintillation counting cocktails; liquid scintillation counting vials, gel dryer.

2.7 Fam20C Activity in Cells

1. Tissue culture components: U2-OS osteosarcoma cells (Control and Fam20C KO). Corning® Costar® six-well cell culture plates; Dialyzed FBS; Phosphate free DMEM. DMEM cell growth medium; Fetal bovine serum (FBS); DPBS; Pen/Strep.

2. [^{32}P]orthophosphate, 10 mCi/mL.

3. pCCF-Fam20C, pCCF-Fam20C D478A, and pcDNAV5-OPN mammalian expression plasmids.

4. Transfection reagent: Fugene 6.

5. Protein G Agarose.

6. DPBS containing 0.4 mM EDTA and 1% NP40.

7. Rabbit anti-V5 antibody (Millipore), mouse anti-V5 antibody (Invitrogen), and mouse M2 anti-Flag antibody (Sigma).

8. Gel electrophoresis and transfer system.

3   Methods

Carry out all procedures at room temperature unless otherwise specified.

3.1   Expression and Purification of Recombinant Fam20B and Fam20C in Insect Cells

1. Insect cell maintenance.

   Cells are grown at 27 °C in a non-humidified shaker with constant shaking at 115 rpm. The cell density is maintained between 0.5×10^6 and 2.5×10^6 cells/mL by splitting and diluting the culture with fresh medium (*see* **Notes 1** and **2**).

2. Subclone Fam20B and Fam20C into the pSMBP2 Vector.

   DNA fragment encoding the kinase domain of Fam20B (residues 32-409) or Fam20C (residues 63-584) are subcloned into the multiple cloning sites of pSMBP2 in frame of the TEV site. This will allow their expression as a fusion protein. The gp67 signal peptide will mediate the secretion of the fusion protein, the 6X His tag facilitates protein purification from the conditioned medium, and MBP aids in the proper folding of the kinases.

   The presence of MBP is essential for the folding and secretion of the Fam20 kinases. Previously, we also used SUMOstar as a fusion tag to enhance their folding [34]. SUMOstar is a mutant of yeast SUMO (Small Ubiquitin-like Modifer) that is resistant to endogenous Ulp1 desumoylase [35]. We found that MBP generally performs better than SUMOstar and increases the secretion of the Fam20 kinases by >2–3 fold.

   N-terminal sequencing analysis of Fam20C purified from the conditioned medium of HEK293T cells revealed that the mature protein starts at Asp93 [16]. However, when Fam20C (93-584) was expressed in insect cells using the above strategy, the resulting fusion protein was found to be extremely resistant to cleavage by the TEV protease, possibly due to the presence of an *N*-linked glycosylation site at Asn101. Fam20C (63-584) behaves well in solution and is similarly active.

3. Generation of the recombinant bacmid.

   Follow the Bac-to-Bac system protocol (Invitrogen) to transform the pSMBP2 plasmids into DH10Bac *E. coli* and transpose the expression cassette into the bacmid. Pick a large white colony and inoculate 10 mL of Luria–Bertani (LB) medium. Grow overnight at 37 °C, and isolate the bacmid DNA using the PureLink HiPure Plasmid DNA miniprep Kit.

   We transform 1 ng of plasmid DNA to 20 μL of MAX Efficiency DH10Bac competent cells, instead of 100 μL as recommended by Invitrogen. Invitrogen also recommends picking ten white colonies and verifying them by re-streaking on fresh growth plates. We find the verification step is usually unnecessary. To reduce cost, the bacmid DNA can also be isolated by the traditional isopropanol precipitation method.

4. Produce and amplify the baculovirus.

   Transfect 1 μg of bacmid DNA to 0.8–1 × 10⁶ healthy Sf9 or Sf21 cells seeded in one well of a six-well plate using 8 μL of Cellfectin or X-tremeGENE 9 transfection reagent. Change to fresh medium after 3–5 h of transfection, and incubate at 27 °C for 72–96 h (*see* **Note 3**).

   Collect the conditioned medium containing the P1 virus, filter through a 0.2 μm syringe filter, and store at 4 °C in the dark.

   To amplify the virus and make P2 viral stock, infect Sf9 or Sf21 culture at 1.5–2 × 10⁶ cells/mL using 1:500 P1–culture volume ratio. Grow the cells in suspension with constant shaking at 115 rpm. Collect the culture supernatant after 72–96 h of infection, filter through a 0.22 μm filter, and store at 4 °C until use. Usually, the titer of the P2 virus is sufficient for protein expression.

   It may be helpful to supplement the viral stock with 2 % fetal bovine serum to prevent aggregation of the viral particles.

5. Protein expression in High Five cells.

   High Five cells are the best for expression and secretion of the Fam20 kinases. Sf21 cells can also be used, but the protein production is at least two- to threefold lower. Sf9 cells cannot secrete the Fam20 proteins.

   Infect suspension cultures of High Five cells at 1.5 × 2 × 10⁶ cells/mL using ~1:200 P2–culture volume ratio (*see* **Note 4**). Harvest the conditioned medium after 48–60 h of infection by spinning down the cells at 1500 × *g* for 15 min at 4 °C.

6. Purification of the Fam20 proteins.

   All protein purification steps were performed at 4 °C. The conditioned medium obtained above is centrifuged again at 8000 x rpm for 30 min, before passing though a 0.22 μm filter. The cleared medium is then concentrated by a cross-flow filtration system. We use two Vivaflow 200 cassettes (cutoff: 30 kDa) connected in parallel. This system allows us to concentrate 2 L of cleared medium to 100 mL in ~4 h. The concentrated sample is subsequently dialyzed against 10 L of TBS buffer (*see* **Note 5**).

   After dialysis, the fusion protein is purified by Ni-NTA affinity method. 2 mL Ni-NTA agarose prewashed with TBS buffer is added to the dialyzed sample, and batch binding is performed for ~2 h with gentle rotation. The Ni-NTA resin is then packed into an empty column, and washed extensively with the wash buffer. Bound proteins are eluted with 10 mL elution buffer and analyzed by SDS-PAGE.

   To remove the 6X His-MBP-fusion tag, the fusion protein is incubated with TEV protease (50:1), and dialyzed against 25 mM Tris pH 8.0, 100 mM NaCl overnight. Untagged Fam20 proteins are further purified using ion-exchange chromatography on a Resource Q column with a 50–500 mM NaCl gradient, followed by size-exclusion chromatography on a Hiload Superdex 200 column (*see* Fig. 2 and **Note 6**).

3.2   *In Vitro* Fam20B Kinase Assay Using Tetra-Ben as the Substrate

1. Activate and equilibrate the Sep-Pak cartridges.

   Insert Sep-Pak cartridges into 50 mL falcon tubes and activate the cartridges with 3 mL methanol followed by equilibration 3× with 2 mL of 0.2 M $(NH_4)_2SO_4$. Let all solvents drain by gravity (*see* **Note 7**).

2. Prepare 100 mM ATP stocks.

   Dissolve ATP in 10 mM HEPES pH 7 and adjust the pH to 7 with NaOH (*see* **Note 8**). Aliquot and store at –80 °C. The 100 mM solution of ATP can then be diluted to 1 mM and [γ-^{32}P]ATP can be added to produce the desired specific radioactivity. We typically use a specific radioactivity of 200 cpm/pmol for Fam20B kinase assays.

3. Prepare Fam20B kinase reaction mix components.

   Determine the number of assay samples including the proper controls (i.e., no kinase control, no substrate control), and prepare the samples accordingly. Each reaction (20 μL) should contain 1× Fam20B kinase buffer (50 mM HEPES, 10 mM $MgCl_2$, 100 mM NaCl, 0.2 % Triton X-100), 5 mM $MnCl_2$, 500 μM [γ-^{32}P]ATP (specific radioactivity ~200 cpm/pmol), and 5 μg/mL Fam20B or Fam20B D309A. The enzyme is diluted in TBS and reactions are typically started by the addition of enzymes.

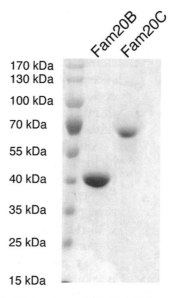

Fig. 2 Purification of Fam20B and Fam20C. SDS-PAGE and Coomassie staining of Fam20B and Fam20C purified from the conditioned medium of Hi5 insect cells

4. Incubate the reactions at 37 °C for 10 min and terminate by adding 100 µL of kinase reaction quenching buffer.

5. Load the solution onto the pre-equilibrated Sep-Pack C18 cartridge and allow the solution to pass through the column bed.

6. Wash the cartridge five times with 2 mL of 0.2 M $(NH_4)_2SO4$ (*see* **Note 9**).

7. Elute the substrate with 1 mL methanol and collect the eluent in a scintillation vial.

8. Add 3 mL of scintillation cocktail and measure the incorporated radioactivity by liquid scintillation counting.

3.3 Purification of Recombinant DCN from the Conditioned Medium of HEK293T Cells

1. Add 50 mL FBS and 5 mL Pen/Strep per 500 mL DMEM. Seed 5 million HEK293T cells in a TripleFlask with 200 mL culture medium. Grow cells in a tissue culture incubator at 37 °C with 5 % CO_2.

2. Transfect cells 24 h after seeding by mixing 25 µg pCCF-DCN in 2.5 mL PBS. Add 75 µL of 1 mg/mL PEI solution. Mix well and incubate at RT for 15 min then add the mixture dropwise to the cells (*see* **Note 10**).

3. Collect the conditioned medium ~48 h later and centrifuge at $13,000 \times g$ for 10 min to clear debris.

4. Add 0.8 mL of washed anti-FLAG M2 affinity gel to the supernatant and rotate the mixture on a roller for 2 h at 4 °C. Carefully pour the solution onto an empty PD-10 column.

5. Wash the resin three times with 10 mL ice-cold TBS/0.1% Triton X-100 and elute the bound Flag tagged DCN with 100 µg/mL FLAG peptide in TBS (3×1 mL).

6. Combine and concentrate the eluted protein using an Amicon centrifugal filter units (30 kDa MWCO). Measure the protein concentration by Bradford and adjust the protein concentration to 1 mg/mL, aliquot, flash freeze in liquid N_2, and store at –80 °C until use.

3.4 In Vitro Fam20B Kinase Assay Using Recombinant DCN as a Substrate

1. For a 20 µL kinase reaction using DCN as a substrate, mix 2 µL of 1 mg/mL Flag-tagged DCN, 1 µL 10× GAG treatment buffer, 1 µL Chon-ABC (1 mU), and 6 µL H_2O. Briefly vortex the tube and incubate at 37 °C for 30 min (*see* **Note 11**). Following incubation, add 4 µL of 5× Fam20B kinase buffer and 2 µL H_2O. Briefly vortex the tube and start the reaction by adding 2 µL of 50 µg/mL Fam20B or Fam20B D309A enzyme. Incubate at 37 °C for 1 h.

2. Terminate the reactions by adding 5 µL of 5× SDS-loading buffer and boil the samples for 10 min. Separate the reaction products by SDS-PAGE and visualize the protein and incorporated radioactivity by Coomassie blue staining and autoradiography, respectively (Fig. 3, *see* **Note 12**).

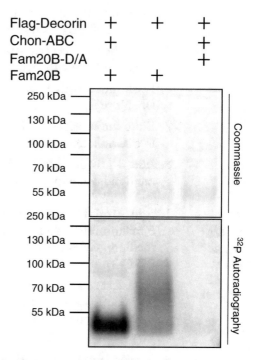

Fig. 3 In vitro Fam20B kinase activity using Flag tagged decorin as the substrate. DCN was purified from the conditioned medium of HEK293T cells and phosphorylated with insect cell purified recombinant Fam20B or the catalytically inactive mutant D309A in the presence of [γ-³²P]ATP. The reactions were treated with or without Chon-ABC and the products were separated by gel electrophoresis and visualized by Coomassie blue staining (*top*) and autoradiography (*bottom*)

3.5 In Cell Fam20B Kinase Assay Using DCN as a Substrate

1. Seed control and Fam20B stable knockdown MRC-5 cells in 10 cm tissue culture dish so that the cells will be 70–80 % confluent ~48 h later.

2. Approximately 48 h after seeding the cells, prepare ³²P cell labeling medium (14 mL phosphate free DMEM, 300 μL dialyzed FBS, 150 μL Pen/Strep, 2 mCi ³²P-orthophosphate).

3. Wash the cells with phosphate free DMEM three times then add 7 mL of ³²P cell labeling medium to each 10 cm dish.

4. Approximately 16 h later, collect the conditioned medium and clear the cell debris by centrifugation at $5000 \times g$ for 5 min. Add protease and phosphatase inhibitor cocktails and concentrate the medium to 2 mL or less using a 30 kDa MWCO Amicon centrifugal filter unit (*see* **Note 13**).

5. Add 10 μL anti-DCN antibody and incubate at 4 °C for 3 h with rotating.

6. Add 15 μL washed Protein G beads and incubate for 1 h at 4 °C with rotating.

7. Wash the beads five times with 2 mL TBS/0.1 % Triton X-100.

8. GAG digestion: Add 10 μL TBS, 2 μL 10× GAG treatment buffer, and 2 μL Chon-ABC and incubate the solution at 37 °C for 30 min.

9. Terminate the reactions by adding 8 μL of 5× SDS loading buffer to the tubes and boil for 10 min.

10. Spin down the beads at $5000 \times g$ for 2 min and use 15 μL of the supernatant for SDS-PAGE and anti-DCN immunoblotting. Compare the amount of DCN from the control and Fam20B knockdown cells by anti-DCN immunoblotting and the amount of ^{32}P incorporation by autoradiography.

3.6 In Vitro Fam20C Peptide Kinase Assays Using β28-40 as the Substrate (See Note 14).

1. Prepare a 10 mM stock solution of β28-40.
   Dissolve the peptide in 10 mM HEPES pH = 7 and adjust the pH to 7 with 5 M NaOH. Aliquot the peptide and store at –20 °C.

2. Prepare Fam20C kinase reaction mix components.
   Determine the number of assay samples including the proper controls (i.e., no kinase control, no substrate control), and prepare the samples accordingly. Each reaction (50 μL) should contain 1× Fam20C kinase buffer (50 mM HEPES pH 7, 0.5 mg/mL BSA), 5 mM MnCl$_2$, 1 mM β(28-40), 100 μM [γ-^{32}P]ATP (specific radioactivity ~500 cpm/pmol), and 1 μg/mL Fam20C or Fam20C D478A. The enzymes are diluted in TBS and reactions are typically started by the addition of [γ-^{32}P]ATP or enzymes. Incubate the reactions at 30 °C for 10–15 min. Also prepare samples to quantify the total radioactivity in the sample. These samples will be spotted onto the P81 filter papers without washing and will be used to determine the specific activity of the reaction mix.

3. Terminate the reactions by spotting 35 μL on 2×2 cm P81 phosphocellulose filter papers and immediately immerse the papers in 300–500 mL of 75 mM phosphoric acid while stirring in a wire mesh basket contained within a beaker.

4. Wash for 10–15 min (*see* Note 15).

5. Discard the phosphoric acid appropriately and add fresh phosphoric acid and continue washing for an additional 30–60 min.

6. Discard the phosphoric acid appropriately and perform a third wash for 10–15 min, followed by a 5 min wash with acetone.

7. Allow the P81 filter papers to air-dry. A hair dryer can be used to reduce the time it takes for the filter papers to dry.

8. Add the filter papers to scintillation vials and monitor the incorporated radioactivity by Cerenkov counting in a Beckman LS 6000IC scintillation counter. If the counts are low, scintillation fluid can be used as well.

**3.7 In Vitro Fam20C
Protein Kinase Assays
Using Recombinant
Osteopontin (OPN)**

1. Recombinant 6X-His tagged osteopontin (residues 17–314) can be expressed and purified from *E. coli* extracts using standard protein expression and purification techniques.

2. Prepare Fam20C kinase reaction mix components.
   Determine the number of assay samples including the proper controls (i.e., no kinase control, no substrate control), and prepare the samples accordingly. Each reaction (20 μL) should contain 50 mM HEPES pH 7, 10 mM $MnCl_2$, 0.25 mg/mL 6X-His osteopontin, 1 μM [γ-^{32}P]ATP (specific radioactivity ~500 cpm/pmol), and 10 μg/mL Fam20C or Fam20C D478A. The enzymes are diluted in TBS and reactions are typically started by the addition of [γ-^{32}P]ATP or enzymes.

3. Incubate the reactions for 2 h at 30 °C.

4. Terminate the reactions by adding EDTA to a final concentration of 15–20 mM, followed by 5× SDS loading buffer and boil the samples for 5 min.

5. Separate the reaction products by SDS-PAGE and visualize the protein and incorporated radioactivity by Coomassie blue staining and autoradiography, respectively (Fig. 4a).

**3.8 Cell Based
Fam20C Kinase Assay
Using OPN
as a Substrate: Gel
Shift Assay**

1. Seed 5×10^5 U2OS cells in 2 mL in a six-well plate format.

2. Approximately 24 h later, mix 1 μL of pCCF-Fam20C or pCCF-Fam20C D478A with 4 μg of pcDNAV5-OPN (*see* **Note 16**).

3. In a separate tube, add 7 μL of FuGENE 6 transfection reagent to 100 μL serum free, Pen/Strep free DMEM.

4. Add 2 μL of the DNA mix and let sit at room temperature for 20–30 min.

5. Add the transfection mixture dropwise to cells and change the medium 4–6 h later.

6. Harvest the conditioned medium 24–48 h later by centrifugation at $750 \times g$ to remove cell debris. Remove the supernatant and perform an addition centrifugation at $10,000 \times g$ for 10 min at 4 °C.

7. Add 1 μL of rabbit anti-V5 antibody (Millipore) to the supernatant and rotate on a nutator for 2 h or overnight at 4 °C.

8. Pull down the antibody–antigen complex by adding 20 μL of a 1:1 slurry of PBS washed Protein G agarose and incubate on a nutator for 1 h.

9. Spin the sample at $1000 \times g$ for 2 min at 4 °C. Remove the supernatant and wash the beads with 1 mL of DPBS. Repeat the wash step for a total of three washes. Boiling the beads in 2× SDS loading buffer can directly elute V5-tagged OPN. Alternatively, the beads can be incubated with

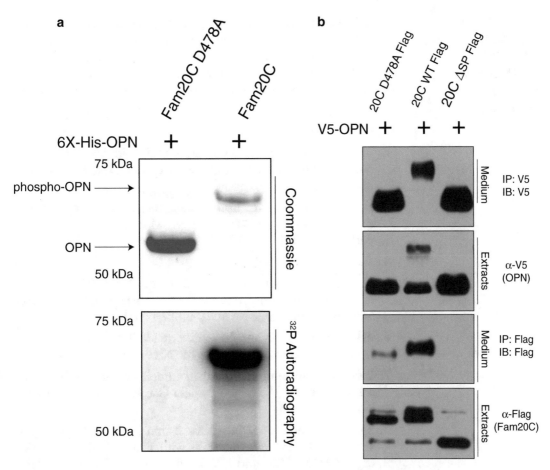

Fig. 4 Phosphorylation of osteopontin in vitro and in cells by Fam20C. (**a**) In vitro incorporation of ^{32}P from [γ-^{32}P] ATP into 6X-His-tagged OPN by Fam20C, or the D478A mutant. Reaction products were separated by SDS-PAGE, visualized by Coomassie Blue staining, and radioactivity was detected by autoradiography. (**b**) Protein immunoblotting of V5 and Flag immunoprecipitates from conditioned medium of U2OS cells co-expressing V5-tagged OPN (V5-OPN) with either WT Fam20C (20C WT Flag), catalytically inactive, D478A Fam20C (20C D478A Flag) or Fam20C lacking the signal peptide (20C ΔSP Flag). Extracts were analyzed for Flag-Fam20C (or mutants) and V5-tagged OPN

λ-phosphatase to remove the phosphate incorporated by Fam20C prior to the addition of SDS loading buffer.

10. Separate the reaction products by SDS-PAGE and visualize the V5-tagged-OPN by protein immunoblotting using mouse anti-V5 antibody (Invitrogen). In parallel experiments, Flag-tagged Fam20C expression can be monitored in the cell extracts or in the conditioned medium by anti-Flag M2 protein immunoblotting (Fig. 4b).

3.9 Cell Based Fam20C Kinase Assay Using OPN as a Substrate: Metabolic Radiolabeling

Not all proteins display a mobility shift when phosphorylated by Fam20C. Therefore, if your protein of interest does not shift when phosphorylated, we recommend metabolically radiolabeling cells with ^{32}P orthophosphate. The proteins can be coexpressed with Fam20C or expressed alone in control and Fam20C depleted cells.

1. Seed 5×10^5 control or Fam20C KO U2OS cells in 2 mL in a six-well plate.

2. Transfect 5 μg of pcDNAV5-osteopontin with 10 μL FuGENE 6 and 100 μL of serum free, PenStrep free DMEM following the protocol outlined above.

3. Forty to forty-eight hours after transfection, replace the medium with phosphate-free DMEM containing 10% (vol/vol) dialyzed FBS and 1 mCi/mL [^{32}P]orthophosphate.

4. Incubate for 6–8 h at 37 °C.

5. Harvest the conditioned medium by centrifugation at $750 \times g$ to remove cell debris. Remove the supernatant and perform an additional centrifugation at $10,000 \times g$ for 10 min at 4 °C.

6. Add 1 μL of rabbit anti-V5 antibody (Millipore) to the supernatant and rotate on a nutator for 2 h or overnight at 4 °C.

7. Pull down the antibody–antigen complex by adding 20 μL of a 1:1 slurry of PBS washed Protein G agarose and incubate on a nutator for 1 h.

8. Spin the sample at $1000 \times g$ for 2 min at 4 °C. Remove the supernatant and wash the beads with 1 mL of DPBS containing 0.4 mM EDTA and 1% NP40. Repeat the wash step for a total of six washes. Boil the beads in 2× SDS loading and separate the proteins by SDS-PAGE. Visualize the V5-tagged-OPN by protein immunoblotting using mouse anti-V5 antibody (Invitrogen). The incorporated radioactivity can be monitored by autoradiography.

4   Notes

1. It is very important that the cell culture volume does not exceed 40% of the maximal flask capacity to keep proper aeration. In fact, we found that the cells grow better when the culture occupies less than 20% of the flask volume.

2. There is an excellent *JoVE* (*Journal of Visualized Experiments*) article that nicely illustrates the general process of insect cell culture and maintenance, generation and amplification of baculovirus, and expression and purification of a recombinant protein [36].

3. If the incubator is non-humidified, it is important to wrap the six-well plate using plastic wrap to reduce evaporation.

4. A pilot small-scale protein expression test can be performed at different virus–culture ratios to determine the optimal amount of virus needed for infection. It might also be helpful to measure the titer of the P2 viral stock by a plaque assay or other methods. In our experience, infection using a 1:200 P2–culture ratio generally works well.

   It is important to add the EDTA to stop the reactions. We have found that highly phosphorylated OPN will begin to precipitate in the presence of $MnCl_2$.

5. It is very important to do this buffer exchange step, since the insect cell medium is not compatible with the Ni-NTA resin.

6. The typical yield is 5–10 mg of Fam20B or Fam20C fusion protein, or 3–6 mg of untagged protein per liter of culture media.

7. A vacuum manifold can be used if it is available.

8. To accurately measure the concentration of ATP, dilute a sample of the stock to 20 μM and monitor the absorbance of the diluted sample at 259 nm. The absorbance of a 20 μM stock of ATP at 259 nm should be 0.31.

9. Use a Geiger counter to frequently monitor the level of radioactivity in the washing eluent to check the progress of washing, and adjust the washing accordingly.

10. Touch the pipet tip to the flask wall and gently pipet solution to the flask to avoid disturbing the cells.

11. If the interested proteoglycan has both chondroitin sulfate and heparin sulfate GAG chains, also add 1 mU each of Heparinase I, II, III.

12. The limitation of the HEK293T expression system is that a significant portion of DCN produced from these cells does not contain the GAG modification.

13. It is important to concentrate the conditioned medium before immunoprecipitation of the secreted endogenous DCN.

14. There is an excellent protocol for performing kinase assays that can also be referred to while performing these assays [37].

15. To reduce background it is important that this wash step does not exceed 15 min.

16. It is important to mix the DNA together prior to adding the transfection reagent to assure the proteins will be expressed in the same cells.

Acknowledgements

This work was supported by a K99/R00 Pathway to Independence Award from the National Institutes of Health (K99DK099254 to V.S.T.) and a Welch Foundation Grant (I-1911 to V.S.T).

References

1. Cohen P (2002) The origins of protein phosphorylation. Nat Cell Biol 4:E127–E130

2. Burnett G, Kennedy EP (1954) The enzymatic phosphorylation of proteins. J Biol Chem 211:969–980

3. Manning G, Whyte DB, Martinez R, Hunter T, Sudarsanam S (2002) The protein kinase complement of the human genome. Science 298:1912–1934

4. Hammarsten O (1883) Zur Frage ob Caseín ein einheitlicher Stoff sei. Hoppe-Seyler's Zeitschrift für Physiologische Chemie 7:227–273

5. Tagliabracci VS, Pinna LA, Dixon JE (2013) Secreted protein kinases. Trends Biochem Sci 38:121–130

6. Yalak G, Vogel V (2012) Extracellular phosphorylation and phosphorylated proteins: not just curiosities but physiologically important. Sci Signal 5:re7

7. Mercier JC, Grosclaude F, Ribadeau-Dumas B (1971) Primary structure of bovine s1 casein. Complete sequence. Eur J Biochem 23:41–51

8. Christensen B, Nielsen MS, Haselmann KF, Petersen TE, Sorensen ES (2005) Post-translationally modified residues of native human osteopontin are located in clusters: identification of 36 phosphorylation and five O-glycosylation sites and their biological implications. Biochem J 390:285–292

9. Moses J, Oldberg A, Cheng F, Fransson LA (1997) Biosynthesis of the proteoglycan decorin--transient 2-phosphorylation of xylose during formation of the trisaccharide linkage region. Eur J Biochem 248:521–526

10. Fransson LA, Silverberg I, Carlstedt I (1985) Structure of the heparan sulfate-protein linkage region. Demonstration of the sequence galactosyl-galactosyl-xylose-2-phosphate. J Biol Chem 260:14722–14726

11. Sreelatha A, Kinch LN, Tagliabracci VS (2015) The secretory pathway kinases. Biochim Biophys Acta 1854:1687–1693

12. Ishikawa HO, Takeuchi H, Haltiwanger RS, Irvine KD (2008) Four-jointed is a Golgi kinase that phosphorylates a subset of cadherin domains. Science 321:401–404

13. Koike T, Izumikawa T, Tamura J, Kitagawa H (2009) FAM20B is a kinase that phosphorylates xylose in the glycosaminoglycan-protein linkage region. Biochem J 421:157–162

14. Wen J, Xiao J, Rahdar M, Choudhury BP, Cui J, Taylor GS, Esko JD, Dixon JE (2014) Xylose phosphorylation functions as a molecular switch to regulate proteoglycan biosynthesis. Proc Natl Acad Sci U S A 111:15723–15728

15. Tagliabracci VS, Engel JL, Wen J, Wiley SE, Worby CA, Kinch LN, Xiao J, Grishin NV, Dixon JE (2012) Secreted kinase phosphorylates extracellular proteins that regulate biomineralization. Science 336:1150–1153

16. Tagliabracci VS, Wiley SE, Guo X, Kinch LN, Durrant E, Wen J, Xiao J, Cui J, Nguyen KB, Engel JL, Coon JJ, Grishin N, Pinna LA, Pagliarini DJ, Dixon JE (2015) A single kinase generates the majority of the secreted phosphoproteome. Cell 161:1619–1632

17. Tagliabracci VS, Engel JL, Wiley SE, Xiao J, Gonzalez DJ, Nidumanda Appaiah H, Koller A, Nizet V, White KE, Dixon JE (2014) Dynamic regulation of FGF23 by Fam20C phosphorylation, GalNAc-T3 glycosylation, and furin proteolysis. Proc Natl Acad Sci U S A 111:5520–5525

18. Ishikawa HO, Xu A, Ogura E, Manning G, Irvine KD (2012) The Raine syndrome protein FAM20C is a Golgi kinase that phosphorylates bio-mineralization proteins. PLoS One 7:e42988

19. Simpson MA, Hsu R, Keir LS, Hao J, Sivapalan G, Ernst LM, Zackai EH, Al-Gazali LI, Hulskamp G, Kingston HM, Prescott TE, Ion A, Patton MA, Murday V, George A, Crosby AH (2007) Mutations in FAM20C are associated with lethal osteosclerotic bone dysplasia (Raine syndrome), highlighting a crucial molecule in bone development. Am J Hum Genet 81:906–912

20. Raine J, Winter RM, Davey A, Tucker SM (1989) Unknown syndrome: microcephaly, hypoplastic nose, exophthalmos, gum hyperplasia, cleft palate, low set ears, and osteosclerosis. J Med Genet 26:786–788

21. Rafaelsen SH, Raeder H, Fagerheim AK, Knappskog P, Carpenter TO, Johansson S, Bjerknes R (2013) Exome sequencing reveals FAM20c mutations associated with FGF23-related hypophosphatemia, dental anomalies and ectopic calcification. J Bone Miner Res 28:1378–1385

22. Simpson MA, Scheuerle A, Hurst J, Patton MA, Stewart H, Crosby AH (2009) Mutations in FAM20C also identified in non-lethal osteosclerotic bone dysplasia. Clin Genet 75:271–276

23. Vogel P, Hansen GM, Read RW, Vance RB, Thiel M, Liu J, Wronski TJ, Smith DD, Jeter-Jones S, Brommage R (2012) Amelogenesis imperfecta and other biomineralization defects in Fam20a and Fam20c null mice. Vet Pathol 49:998–1017

24. Wang X, Wang S, Li C, Gao T, Liu Y, Rangiani A, Sun Y, Hao J, George A, Lu Y, Groppe J,

Yuan B, Feng JQ, Qin C (2012) Inactivation of a novel FGF23 regulator, FAM20C, leads to hypophosphatemic rickets in mice. PLoS Genet 8:e1002708

25. Cui J, Xiao J, Tagliabracci VS, Wen J, Rahdar M, Dixon JE (2015) A secretory kinase complex regulates extracellular protein phosphorylation. Elife 4:e06120

26. Wang SK, Aref P, Hu Y, Milkovich RN, Simmer JP, El-Khateeb M, Daggag H, Baqain ZH, Hu JC (2013) FAM20A mutations can cause enamel-renal syndrome (ERS). PLoS Genet 9:e1003302

27. O'Sullivan J, Bitu CC, Daly SB, Urquhart JE, Barron MJ, Bhaskar SS, Martelli-Junior H, dos Santos Neto PE, Mansilla MA, Murray JC, Coletta RD, Black GC, Dixon MJ (2011) Whole-exome sequencing identifies FAM20A mutations as a cause of amelogenesis imperfecta and gingival hyperplasia syndrome. Am J Hum Genet 88:616–620

28. Bordoli MR, Yum J, Breitkopf SB, Thon JN, Italiano JE Jr, Xiao J, Worby C, Wong SK, Lin G, Edenius M, Keller TL, Asara JM, Dixon JE, Yeo CY, Whitman M (2014) A secreted tyrosine kinase acts in the extracellular environment. Cell 158:1033–1044

29. Dudkiewicz M, Lenart A, Pawlowski K (2013) A novel predicted calcium-regulated kinase family implicated in neurological disorders. PLoS One 8:e66427

30. Morrow EM, Yoo SY, Flavell SW, Kim TK, Lin Y, Hill RS, Mukaddes NM, Balkhy S, Gascon G, Hashmi A, Al-Saad S, Ware J, Joseph RM, Greenblatt R, Gleason D, Ertelt JA, Apse KA, Bodell A, Partlow JN, Barry B, Yao H, Markianos K, Ferland RJ, Greenberg ME, Walsh CA (2008) Identifying autism loci and genes by tracing recent shared ancestry. Science 321:218–223

31. Yoshida-Moriguchi T, Willer T, Anderson ME, Venzke D, Whyte T, Muntoni F, Lee H, Nelson SF, Yu L, Campbell KP (2013) SGK196 is a glycosylation-specific O-mannose kinase required for dystroglycan function. Science 341:896–899

32. Jae LT, Raaben M, Riemersma M, van Beusekom E, Blomen VA, Velds A, Kerkhoven RM, Carette JE, Topaloglu H, Meinecke P, Wessels MW, Lefeber DJ, Whelan SP, van Bokhoven H, Brummelkamp TR (2013) Deciphering the glycosylome of dystroglycanopathies using haploid screens for lassa virus entry. Science 340:479–483

33. von Renesse A, Petkova MV, Lutzkendorf S, Heinemeyer J, Gill E, Hubner C, von Moers A, Stenzel W, Schuelke M (2014) POMK mutation in a family with congenital muscular dystrophy with merosin deficiency, hypomyelination, mild hearing deficit and intellectual disability. J Med Genet 51:275–282

34. Xiao J, Tagliabracci VS, Wen J, Kim SA, Dixon JE (2013) Crystal structure of the Golgi casein kinase. Proc Natl Acad Sci U S A 110:10574–10579

35. Liu L, Spurrier J, Butt TR, Strickler JE (2008) Enhanced protein expression in the baculovirus/insect cell system using engineered SUMO fusions. Protein Expr Purif 62:21–28

36. Yates LA, Gilbert RJ (2014) Efficient production and purification of recombinant murine kindlin-3 from insect cells for biophysical studies. J Vis Exp 85:51206

37. Hastie CJ, McLauchlan HJ, Cohen P (2006) Assay of protein kinases using radiolabeled ATP: a protocol. Nat Protoc 1:968–971

INDEX

William J. Brown (ed.), *The Golgi Complex: Methods and Protocols*, Methods in Molecular Biology, vol. 1496,
DOI 10.1007/978-1-4939-6463-5, © Springer Science+Business Media New York 2016

Printed in the United States
By Bookmasters